Mass Spectrometry - Future Perceptions and Applications

Edited by Ganesh Shamrao Kamble

Published in London, United Kingdom

IntechOpen

Supporting open minds since 2005

Mass Spectrometry - Future Perceptions and Applications
http://dx.doi.org/10.5772/intechopen.78510
Edited by Ganesh Shamrao Kamble

Contributors
Vijayabhaskar Veeravalli, Giulio Tarro, Alessandra Rossi, Moreno Paolini, Xianquan Zhan, Tian Zhou, Oscar Núñez, Guillem Campmajó, Nerea Núñez, Najeeb Punnakayathil, Gunawan Witjaksono, Sagir Alva

First published in London, United Kingdom, 2019 by IntechOpen
IntechOpen is the global imprint of INTECHOPEN LIMITED, registered in England and Wales, registration number: 11086078, 7th floor, 10 Lower Thames Street, London, EC3R 6AF, United Kingdom
Printed in Croatia

British Library Cataloguing-in-Publication Data
A catalogue record for this book is available from the British Library

Additional hard and PDF copies can be obtained from orders@intechopen.com

Mass Spectrometry - Future Perceptions and Applications
Edited by Ganesh Shamrao Kamble
p. cm.
Print ISBN 978-1-83962-756-9
Online ISBN 978-1-83962-757-6
eBook (PDF) ISBN 978-1-83962-758-3

For EU product safety concerns:
IN TECH d.o.o., Prolaz Marije Krucifikse Kozulić 3, 51000 Rijeka, Croatia, info@intechopen.com or visit our website at intechopen.com.

Meet the editor

Dr. Ganesh S. Kamble completed his PhD in Chemistry from the Department of Chemistry, Shivaji University Kolhapur, India, in 2011. Most of his research focuses on analytical and material chemistry. After his PhD, he became an assistant professor at the Department of Engineering Chemistry, Kolhapur Institute of Technology's College of Engineering (Autonomous), Kolhapur, Maharashtra, India. He is a life member of the Association of Separation Scientists and Technologists, Bhaba Atomic Research Centre, Mumbai, India. He is a reviewer of peer-reviewed international journals and many other publications. He was awarded "Summer Research Fellow-2013" by the Indian Academy of Sciences, Bangalore. He was also awarded a Postdoctoral Research Fellowship by the Ministry of Science and Technology, P.R. China, and completed his postdoc research at the Department of Chemistry, National Tsing Hua University, Taiwan.

Contents

Preface

Among the analytical techniques, mass spectrometry is one of the commanding analytical tools, which can be used for illuminating the structure and chemical properties of different molecules, identifying unknown compounds within a sample, and quantifying the concentration of known materials, etc. In recent years it has been used as a powerful analytical tool in biological chemistry, specifically in proteins, oligonucleotides, oligosaccharides, and lipids. Apart from this, it is also used for food toxicity, cancer cells, analysis of availability of minerals in food, and pharmaceutical drug analysis.

Today's world faces many challenges in the fields of life sciences, metabolism, food, health, pharmaceuticals, as well as the environment. It has been widely illustrated that mass spectrometric techniques, including MALDI, tandem mass spectrometry, ion cyclotron resonance and Fourier transform mass spectrometry, time-of-flight secondary ion mass spectrophotometry and Fourier transform mass spectrometry, etc. have shown a significant role in major challenges and innovations in both technology and practical applications. Because of its characterization, mass spectrometry is being used in unlimited potential applications in cancer cells, material characterization, marine biology, organic synthesis, and food analysis.

As editor of this book, I wish to express my sincere thanks to all authors, scientists, researchers, and publishers who generously contributed their research work. My thanks are extended to my PhD research guide Professor M.A. Anuse, Department of Chemistry, Shivaji University Kolhapur, for his constant support. I am also indebted to my postdoctoral research supervisor Professor Yong-Chien Ling, Department of Chemistry, National Tsing Hua University, Taiwan, from whom I took inspiration to complete tasks on mass spectrometry.

In spite of all efforts, mistakes or uncertain reports may remain within the text. Mistakes are human traits, but they do not undervalue the scientific or educational worth of a book. Therefore, it is requested that errors are reported to me so that corrections can be made in future editions. Hopefully, *Mass Spectrometry—Future Perceptions and Applications* will introduce many facets of mass spectrometry and give comprehensive information on sustained progressive research and applications.

Dr. Ganesh S. Kamble
Kolhapur Institute of Technology's College of Engineering (Autonomous),
Kolhapur, India

Section 1

Mass Spectrometer and Techniques

Chapter 1

The Role of Liquid Chromatography-Mass Spectrometry in Food Integrity and Authenticity

Guillem Campmajó, Nerea Núñez and Oscar Núñez

Abstract

Liquid chromatography coupled to mass spectrometry (LC-MS), tandem mass spectrometry (LC-MS/MS), and high resolution mass spectrometry (LC-HRMS) today are among the most common techniques to guarantee food integrity and authenticity. Targeted approaches, where a family of characteristic bioactive substances in the analyzed food products are monitored, are a common practice to ensure food authenticity regarding the production region since bioactive substances content and distribution in food depend on multiple parameters such as climate conditions, water resources, agrochemical practices, etc. On the other hand, non-targeted approaches, such as metabolomic fingerprinting, are a common practice where a huge number of spectral detected variables in the analyzed foods are monitored. In both approaches, characteristic patterns are searched among the analyzed food products by means of statistical chemometric methods to address food characterization, classification, and authentication. In the present chapter, the role of LC-MS in combination with chemometrics to guarantee food integrity and authenticity will be discussed. Coverage of all kinds of applications is beyond the scope of the present contribution, so we will focus on the most relevant applications published in the last years by addressing the most interesting examples and important aspects in the food authenticity field.

Keywords: food authenticity, liquid chromatography, mass spectrometry, high resolution mass spectrometry

1. Introduction

Food products are very complex mixtures constituted by a great variability of naturally occurring compounds such as lipids, carbohydrates, proteins, vitamins, organic acids, and volatile organic compounds, among others. Moreover, they can also contain many other substances coming from agrochemical treatments and technological processes, or even migrating from the materials employed in food packaging, which sometimes are contaminants.

Food manufacturers, researchers, and society in general are also becoming very interested in the quality of food products, not only from the nutritional point of view but also in relation to food safety issues or regarding the presence of bioactive substances with beneficial properties for consumers (functional foods, nutraceuticals, etc.). Aspects related to the cultivation production region of natural food products (fruits, vegetables, etc.), as well as the cultivation techniques employed, begin to be also of great interest to the final consumers, giving rise to the consideration of the protected designations of origin (PDO) of natural foodstuffs as important food quality attributes.

Nowadays, the food supply production is worldwide distributed and consequently a globalized issue. Although international and local regulatory bodies have established important rules in the labeling of food products, in general, it is often almost impossible to know the real origin of most of the components of a given food product, especially those that have been processed. Within this context, considering the complexity of the food chain and that many players are involved between production and consumption; food manipulation and adulteration practices are raising because it is in fact much easier to conduct fraud without being easily detected. For example, Moore et al. collected information from published articles in scholarly journals and general media, organized it into a database, and reviewed and analyzed the data to identify trends within food ingredient fraud practices from 1980 to 2010 [1]. They observed that olive oil, milk, honey, and saffron were the most common targets for adulteration reported in scholarly journals and potentially harmful issues identified include spices diluted with lead chromate and lead tetraoxide, substitution of Chinese star anise with toxic Japanese star anise, and melamine adulteration of high protein content foods.

Food adulteration practices have a long history and dates back to times when trading began. In general, food adulteration is carried out to increase volume, to mask the presence of inferior quality components and to replace the authentic substances for the seller's economic gain. However, it must be considered that the deliberate adulteration of food and its misrepresentation to deceive final consumers is illegal worldwide, having not only economic consequences, but also representing important health issues when prohibited substances are added to deceive the organoleptic properties of the final food product or when the adulterant can produce allergy episodes. Thus, the development of new analytical methodologies to guarantee food integrity and authenticity is required, also considering that food adulteration has become increasingly sophisticated, often being specially designed to avoid detection through routine analysis approaches.

The analysis of food products is difficult not only because of the complexity and diversity of sample matrices but also due to the great variability of compounds that can be present. In addition, food components differ in polarity, structures, as well as in concentration levels, going from components at grams per kilogram level to those found at trace level concentrations (low μg/kg, ng/kg, etc.). These are important aspects to consider when selecting the analytical approach to employ. Sample treatment and sample extraction procedures, separation and determination approaches, and identification and confirmation strategies need to be considered simultaneously when addressing the development of an analytical method in food integrity and authenticity analyses. Nowadays, liquid chromatography coupled to mass spectrometry (LC-MS) or to tandem mass spectrometry (LC-MS/MS) is among the most effective analytical techniques for the structural characterization and analysis of food products. The appearance of ultra-high performance liquid chromatography (UHPLC) methodologies, either using sub-2 μm particle packed columns or porous-shell columns (with sub-3 μm superficially porous particles), opened up new possibilities to achieve high throughput chromatographic analytical separations, 5- to 10-fold faster than with conventional LC methodologies, while keeping or even improving chromatographic resolutions [2]. The use of liquid chromatography coupled to high resolution mass spectrometry (LC-HRMS) and accurate mass measurements have recently gained huge popularity due to the great ability of these methodologies to provide more comprehensive information regarding the exact molecular mass, elemental composition, and detailed molecular structure of a given compound. In comparison to classical low resolution mass spectrometry (LRMS) techniques, HRMS allows to differentiate isobaric compounds (substances with the same nominal mass-to-charge ratio but different elemental compositions). Moreover, the high resolution attainable with HRMS favors the simplification of sample treatment and preparation procedures, leading to faster analytical methodologies with less and simple sample manipulation. HRMS allows to perform both screening and quantitation in a single run, including targeted, suspect, and non-targeted analyses. Another important advantage of HRMS, especially when data is stored in full-scan mode, is the possibility of later stage retrospective analysis, allowing the identification and determination of new unknown or suspected compounds in a previously analyzed food sample.

An important aspect in food products, especially those of plant origin, is that the presence, distribution, and content of many bioactive substances is related to many food features such as the variety and species of the products, the degree of maturation in the fruits and vegetables employed, the geographical production areas, the growing and manufacturing practices used, etc. A similar consideration can be mentioned for food products of animal origin, where many substances present in the final product will be related to the animal species, the farming practices employed, the animal stress, etc. Therefore, food chemical profiling, for instance of amino acids, biogenic amines, alcohols, aldehydes, esters, acids, terpenes, polyphenols, etc., can be exploited as sample data descriptors to achieve the characterization, classification, and authentication of food products.

Regarding chemical profiling in food integrity and authenticity by LC-MS and LC-HRMS methodologies, two main approaches are typically employed: targeted and non-targeted analyses. Targeted approaches can be performed by both LC-MS(/MS) and LC-HRMS techniques and are based on the specific determination of a given group of known selected chemicals, or a group of chemicals belonging to the same family or with a similar structural feature. The concentrations (or peak signals) of these targeted compounds are then used as food features (markers) to address food integrity and authenticity. This approach requires, in general, a previous quantitation step using standards for each targeted component. However, when dealing with food products, which as previously commented are very complex matrices, the quantitation of some chemicals may be a difficult task, especially due to the possibility of unknown interfering compounds. In contrast, non-targeted approaches (based on metabolomic fingerprinting) are mainly employed with LC-HRMS techniques. These fingerprinting approaches are based on untargeted analysis of instrumental responses without assuming any previous knowledge of relevant or irrelevant food components. In the case of LC-HRMS, food sample fingerprinting information consists, in general, of peak intensity values recorded as a function of *m/z* and retention times [3].

Due to the complexity of food sample matrices and the variability of chemical components that can be present, the amount of chemical data that can be extracted, especially when dealing with non-targeted LC-HRMS fingerprinting approaches, is huge. As a consequence, in order to extract (bio)chemical information from the sample data sets able to characterize, classify and authenticate food products, chemometric data treatment methodologies are necessary. Multivariate methods such as principal component analysis (PCA) and partial least squares-discriminant analysis (PLS-DA) are among the most employed chemometric methods for exploratory and classification purposes in food integrity and authenticity [4].

In the next sections, several examples dealing with targeted and non-targeted strategies based on LC-MS(/MS) and LC-HRMS methodologies, in combination with chemometrics, to guarantee food integrity and authenticity will be addressed.

2. Targeted approaches

2.1 LC-MS(/MS) methodologies

LC-MS and LC-MS/MS are among the most common techniques used in the literature to obtain qualitative, quantitative, and structural information in the determination of low molecular weight compounds in a great variety of sample matrices, including foodstuffs. The low sensitivity typically achieved when LRMS is employed, especially with some analyzers such as triple quadrupole (QqQ) and ion trap (IT) instruments, makes them ideal to be employed when targeted approaches are intended. This strategy is based on the specific determination of a given group of compounds (i.e., some selected chemicals, a group of chemicals belonging to the same family, etc.) that can then be used as biomarkers to address food integrity and authenticity. Although this approach typically requires the quantitation of these chemicals by using adequate standards for each targeted component, in some cases targeted profiling is also possible by means of employing only the peak area signal of a given set of compounds, without the requirement of knowing the concentration values.

Polyphenols, aromatic secondary metabolites ubiquitously spread through the plant kingdom, are among the most common biomarkers employed to address food integrity and authenticity when targeted LC-MS(/MS) methodologies are employed in the analysis of plant-related foodstuffs [5] and some selected applications found in the literature are summarized in **Table 1**.

As can be seen in the table, reversed-phase liquid chromatography (RPLC), mainly employing C18 columns [6–12] and gradient elution with an acidified aqueous solution and methanol or acetonitrile as mobile phase components, is usually proposed. For example, Seraglio et al. [7] described the development of a reproducible and sensitive method for the simultaneous determination of 32 phenolic compounds in bracatinga (*Mimosa scabrella* Bentham) honeydew honey samples using HPLC-ESI-MS/MS. The separation was performed with a C18 reversed-phase column in less than 17 min, using gradient elution with water and acetonitrile, both acidified with 0.1% formic acid. Other stationary phases have also been proposed for the separation of polyphenols in food products. For instance, Alakolanga et al. [13] described the use of a C18 amide reversed-phase column for the determination of 35 phenolic compounds in fruits of *Flacourtia indica* (Burm. F.) Merr. and *Flacourtia inermis* Roxb trees. Due to the high number of compounds and the complexity of the sample matrix, a gradient elution program of 80 min was employed. In another application, a fluorinated porous-shell column (Ascentis Express F5) was proposed for the determination of 15 polyphenolic compounds in *Passiflora subpeltata* fruit pulp with a 38 min gradient elution program [14].

Regarding the ionization of polyphenols, electrospray (ESI) in negative mode [10–13], positive mode [9], or exploring both positive and negative modes [6–8, 14] is generally employed. However, other atmospheric pressure ionization (API) sources have also been described in the literature for the determination of polyphenols in food characterization and authentication. For example, Parets et al. [12] compared the use of ESI, atmospheric pressure chemical ionization (APCI), and dopant-assisted atmospheric pressure photoionization using four organic solvents as dopants (toluene, acetone, chlorobenzene, and anisole) for

Compounds (sample)	Chromatographic separation and mass spectrometry	Data analysis	Ref.
Polyphenols (pomegranate)	Ascentis Express C18 column (150 × 3.0 mm, 2.7 μm) Gradient elution (0.4 mL·min^{-1}): (A) water with 2% formic acid (B) methanol:water 90:10 (v/v) with 0.5% formic acid H-ESI (±) IT (full-scan 100–1500 *m/z* and MS^2 product ion scan mode 50–1500 *m/z*)	–	[6]
Polyphenols (*Passiflora subpeltata* fruit)	Ascentis Express F5 (150 × 2.1 mm, 2.7 μm) Gradient elution (0.2 mL·min^{-1}): (A) water with 0.1% formic acid (B) acetonitrile H-ESI (±) QqQ (MRM acquisition mode)	–	[14]
Phenolic compounds (honey)	VENUSIL C18 column (100 × 2.1 mm, 3 μm) Gradient elution (0.3 mL·min^{-1}): (A) water with 0.1% formic acid (B) acetonitrile with 0.1% formic acid H-ESI (±) Q-TRAP (MRM acquisition mode)	–	[7]
Phenolic compounds (*Flacourtia indica* and *Flacourtia inermis* fruit)	C18 amide column (250 × 3 mm, 5 μm) Gradient elution (0.5 mL·min^{-1}): (A) water with 0.05% formic acid (B) methanol ESI (−) IT (full-scan and auto-MS^n mode)	–	[13]
Phenolic compounds (artichoke, garlic and spinach)	Zorbax Eclipse Plus C18 column (100 × 2.1 mm, 1.8 μm) Gradient elution (0.2 mL·min^{-1}): (A) methanol (B) water with 0.1% formic acid and 30 mM of ammonium acetate H-ESI (±) QqQ (MRM acquisition mode)	–	[8]
Phenolic compounds (berries)	Wakosil C18 column (150 × 4.6 mm, 5 μm) Gradient elution (1 mL·min^{-1}): (A) water with 0.1% formic acid (B) acetonitrile with 0.1% formic acid H-ESI (+) Quadrupole MS (full-scan mode 100–800 *m/z*)	–	[9]
Phenolic compounds (tomato fruits)	BEH Shield RP18 column (150 × 1 mm, 1.7 μm) Gradient elution (0.13 mL·min^{-1}): (A) water:acetonitrile 95:5 (v/v) with 0.1% formic acid (B) water:acetonitrile 40:60 (v/v) with 0.1% formic acid H-ESI (−) QqQ (full-scan and product ion scan mode)	ANOVA	[10]
Polyphenols (fruit extracts)	Kinetex C18 (100 × 4.6 mm, 2.6 μm) Gradient elution (1 mL·min^{-1}): (A) water with 0.1% formic acid (B) methanol H-ESI (−) QqQ (MRM acquisition mode)	PCA	[11]
Polyphenols (cranberry-based pharmaceutical preparations and natural extracts)	Hypersil Gold C18 column (50 × 2.1 mm, 1.9 μm) Gradient elution (0.285 mL·min^{-1}): (A) water with 0.1% formic acid (B) methanol H-ESI (−)/APCI (−)/APPI (−) QqQ (MRM acquisition mode)	PCA	[12]

Heated-electrospray ionization (H-ESI), multiple reaction monitoring (MRM), quadrupole-ion trap (Q-TRAP), electrospray ionization (ESI), analysis of variance (ANOVA), atmospheric pressure chemical ionization (APCI), atmospheric pressure photoionization (APPI).

Table 1.
Selected targeted LC-MS(/MS) methods using polyphenols as biomarkers to address food integrity and authenticity.

the determination of 29 polyphenols in grape- and cranberry-based fruit extracts and cranberry-based pharmaceutical preparations. ESI and acetone-assisted APPI showed a good performance for the ionization step of the targeted polyphenolic compounds, providing good sensitivity for most of the analyzed polyphenols. However, when addressing the classification and authentication of the analyzed extracts, the authors described that results obtained by UHPLC-APPI-MS/MS were more satisfactory and the discrimination of the sample classes was excellent in comparison to UHPLC-ESI-MS/MS, attributing this behavior to the higher robustness of APPI source in the presence of matrix effects.

Quadrupole MS, QqQ, and IT instruments are the most employed for the LC-MS(/MS) determination of polyphenols in food integrity and authenticity. Regarding the acquisition mode, full-scan and product-ion scan modes are typically employed with IT instruments, while MRM acquisition mode is applied with QqQ instruments due to the sensitivity improvement observed in comparison with product-ion scan in this kind of instruments. Nevertheless, some authors are also proposing the use of MRM acquisition mode with Q-TRAP instruments [7], although no special improvement in sensitivity is described. Regarding the use of MS^2, several acquisition strategies can be found in the literature. For instance, Brighenti et al. [6] proposed the use of the SmartFrag function of the IT mass analyzer to ensure that every precursor ion receives the appropriate collision energy in order to obtain adequate product-ion scan spectra with the better fragmentation possible.

In order to address food integrity and authenticity, the comparison of data obtained from different sample matrices is required. Therefore, chemometric methodologies, such as PCA, that allow the comparison of multiple variables play an important role in this aspect. However, several works are addressing food authenticity directly by comparison of targeted bioactive substances' content, without the requirement of employing any chemometric strategy. This is the case, for example, of the work described by Ribas et al. [10] that showed significant differences in the phenolic content of three Spanish tomato varieties depending on the cultivar variety ("Caramba," "Montserra," and "Pera de Girona").

Even though the concentration data of some targeted bioactive substances may allow to directly differentiate some food attributes, as previously commented, this data could also be subjected to chemometric methods to address food integrity and authenticity issues. For instance, Puigventós et al. [11] describes the use of LC-ESI-MS/MS method for the determination of 26 polyphenolic compounds in fruit-based products and fruit-based pharmaceutical preparations. The polyphenolic content was then employed as chemical descriptors to achieve sample classification and authentication by means of PCA. As an example, **Figure 1** shows the PCA plot of scores (a) and plot of loadings (b) for the analyzed samples.

As shown in the plot of scores (**Figure 1a**), grape and cranberry products appeared in different zones so that PCA was basically able to distinguish among the two fruits or origins, allowing the authentication of fruit-based extracts. In particular, grape and related samples were located to the top-left part of the graph. In contrast, cranberry samples were mainly spread out on the bottom area. Regarding the plot of loadings (**Figure 1b**), it was found that gallic acid and polydatin were characteristic of grape-related samples so they were present in higher levels in this class of products. In contrast, analytes located to the right part of PC1 such as sinapic, ferulic, *p*-coumaric and chlorogenic acids, as well as quercitrin, were comparatively more abundant in cranberry products.

Therefore, the chemometric analysis of targeted bioactive substance contents in food products could give an idea of the more discriminant chemical descriptors of a given sample, allowing the proposal of future biomarkers to address food integrity and authenticity.

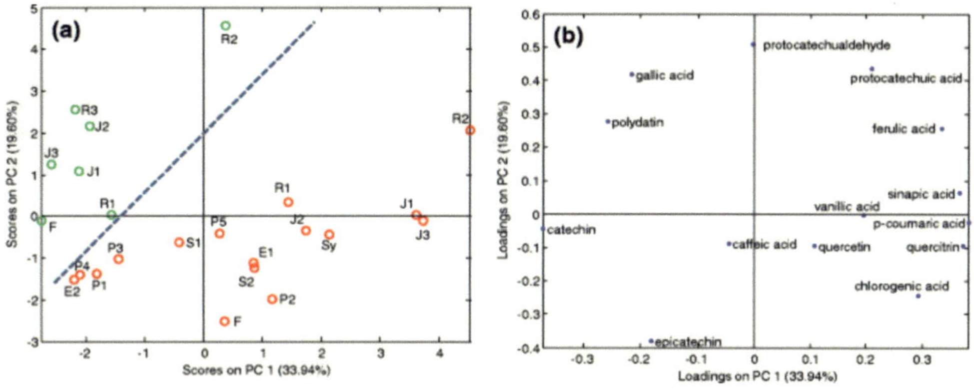

Figure 1.
PCA results using normalized concentrations as analytical data obtained using LC-ESI-MS/MS polyphenolic profiles. (a) Score plot of PC1 vs. PC2 that shows the separation among grapes-based samples (green circles) and cranberry-based samples (red circles). (b) Loading plot PC1 vs. PC2 that shows the polyphenolic content of among grapes- and cranberry-based samples. Reproduced with permission from Ref. [11]. Copyright (2015) Springer.

2.2 LC-HRMS methodologies

Even though LC-MS and LC-MS/MS have proved to be useful techniques in some food authenticity and integrity applications, as previously described, sometimes a more sensitive and selective technique, such as LC-HRMS, is needed mainly due to the complexity of food sample matrices and the huge variability on bioactive compounds, with different structures and physicochemical properties, that they contain [5]. HRMS and accurate mass measurements are emerging as one of the best options for the analysis of food samples in order to guarantee the unequivocal determination of the elemental composition of a target compound, which allows its distinction of other co-eluting isobaric compounds. There are mainly four types of HRMS instruments: magnetic sector, time-of-flight (TOF), Orbitrap, and Fourier transform ion cyclotron resonance (FT-ICR) instruments, being TOF and Orbitrap, as well as some of their hybrid configurations with quadrupole or IT analyzers, the most frequently employed in combination with LC techniques. In general, TOF instruments present a resolution (instrument's ability to measure the mass of two closely related ions precisely) of approximately 10,000–40,000 FWHM (full width at half-maximum) with accuracies in the mass determination of 1–5 ppm. In contrast, the resolution of Orbitrap instruments is in the range of 10,000–140,000 FWHM (or even higher) with 1–2 ppm mass accuracy (for comparison, conventional quadrupole MS instruments show a resolution of 1000 FWHM and accuracies of 500 ppm) [5]. Recent advances in both LC-TOF-MS and LC-Orbitrap-MS methods have reduced instruments costs, make the analysis more simple, and have considerably improved accuracy, offering today bench-top instrumentation that is amenable to screening and identification of a great variety of compounds in food matrices, not only for targeted ones, but also for non-target or unknown chemicals [5].

In this section, the use of targeted LC-HRMS methodologies in order to address the food integrity and authenticity issue will be discussed. **Table 2** summarizes some selected applications described in the literature employing targeted LC-HRMS methodologies in food integrity and authenticity.

As can be seen in **Table 2**, and in line with previously commented targeted LC-MS and LC-MS/MS methodologies, polyphenols are ubiquitously used as biomarkers in targeted LC-HRMS approaches [15–19], whether considering a specific polyphenolic class or a wider selection. However, polyphenols are not always the best choice to solve the analytical problem even when plant-related food products are addressed, and therefore some other compounds can be employed. For instance, Megías-Pérez et al. [20] used the determination of low molecular weight

Compounds (sample)	Chromatographic separation and mass spectrometry	Data analysis	Ref.
Kaempferol derivatives (saffron)	Ascentis Express Fused-core C18 column (100 × 2.1 mm, 2.7 μm) Gradient elution (0.4 mL·min^{-1}): A) water with 0.1% formic acid (B) acetonitrile with 0.1% formic acid H-ESI (−) Q-TOF (full-scan mode 100–1700 *m/z*)	–	[15]
Polyphenols (kiwifruit juice)	Waters XTerra MS C18 column (250 × 4.6 mm, 5 μm) Gradient elution (0.8 mL·min^{-1}): (A) water with 0.5% acetic acid (B) water:acetonitrile 1:1 (v/v) with 0.5% acetic acid H-ESI (−) Q-TOF (full-scan mode 50–1500 *m/z*)	ANOVA, PCA and SLDA	[16]
Polyphenols (berry fruit juice)	Phenomenex C18 column (100 × 2.1 mm, 2.6 μm) Gradient elution (0.3 mL·min^{-1}): (A) water with 0.1% formic acid (B) methanol with 0.1% formic acid H-ESI (±) Q-TOF (full-scan mode 50–1000 *m/z*)	PCA-DA and OPLS-DA	[17]
Polyphenols (red spice paprika)	Syncronis C18 column (100 × 2.1 mm, 1.7 μm) Gradient elution (0.25 mL·min^{-1}): (A) water with 0.01% acetic acid (B) acetonitrile H-ESI (−) LTQ-Orbitrap (full-scan mode 100–1000 *m/z*)	PCA	[18]
Low molecular weight carbohydrates (cocoa beans)	BEH X-Bridge Amide column (150 × 4.6 mm, 3.5 μm) Gradient elution (0.4 mL·min^{-1}): (A) water with 0.1% ammonium hydroxide (B) acetonitrile with 0.1% ammonium hydroxide H-ESI (+) TOF (full-scan mode 50–1200 *m/z*)	ANOVA, PCA and PLS-DA	[20]
Polyphenols (cranberry-based extracts)	Ascentis Express C18 column (150 × 2.1 mm, 2.7 μm) Gradient elution (0.3 mL·min^{-1}): (A) water with 0.1% formic acid (B) acetonitrile with 0.1% formic acid H-ESI (−) Q-Orbitrap (full-scan mode 100–1500 *m/z*)	PCA and PLS	[19]
Small bioactive lipids (rice)	Acquity UPLC BEH C18 column (50 × 2.1 mm, 1.7 μm) Gradient elution (0.5 mL·min^{-1}): (A) water with 10 mM ammonium hydroxide (B) acetonitrile:isopropanol 90:10 (v/v) H-ESI (−) Q-TOF (full-scan mode 50–1200 *m/z*)	PCA and OPLS-DA	[21]

Stepwise linear discriminant analysis (SLDA), principal component analysis-discriminant analysis (PCA-DA), orthogonal partial least-squares discriminant analysis (OPLS-DA), partial least-squares regression (PLS).

Table 2.
Recent advances of targeted LC-HRMS methodologies in food integrity and authenticity.

carbohydrates for the classification of cocoa beans from different origins and status of fermentation, whereas Zhu et al. [21] studied the presence of small bioactive lipids as markers to differentiate among diverse varieties of rice.

Regarding the chromatographic separation, almost all the works described in the literature propose the use of C18 stationary phase columns [15–19, 21]. In fact, Guijarro-Díez et al. [15] tested and compared a C18 and a cyano column, both having the same size and particle diameter, for the chromatographic separation of five kaempferol derivatives and geniposide in the analysis of saffron samples, obtaining a better resolution and peak efficiency when using the C18 column. Alternatively, the separation of polar compounds, such as low molecular weight carbohydrates, can be improved and optimized by using hydrophilic interaction liquid chromatography (HILIC) columns as it offers ample chromatographic resolution [20].

As shown in **Table 2**, H-ESI has been established as the most common option for the ionization step whether positive mode [20], negative mode [15, 16, 18, 19, 21] or both [17] are employed. In order to ensure a good ionization of the targeted compounds (avoiding in-source fragmentation), Guijarro-Díez et al. [15] studied both positive and negative ionization modes, as well as different mobile phase compositions, looking for the highest signal-to-noise (S/N) ratio in the determination of kaempferol derivatives in saffron samples. Negative ESI mode and the addition of 0.1% formic acid to the mobile phase showed to be the best option, being a general trend described in the literature for the determination of polyphenolic compounds [5].

Among all the range of mass analyzers, TOF and Orbitrap-based mass spectrometer technologies are usually employed in these type of targeted studies, especially hybrid instruments such as quadrupole-time-of-flight (Q-TOF) [15–17, 21], quadrupole-Orbitrap (Q-Orbitrap) [19], and linear trap quadrupole-Orbitrap (LTQ-Orbitrap) [18] configurations. The main advantage of these types of instruments in front of single HRMS analyzers is the possibility to make data dependent MS/MS experiments. These acquisition modes provide interesting spectral information that compared with online-databases and with the obtained accurate mass measurements can improve the tentative identification and confirmation of a given targeted compound. It should be mentioned that only when pure standards are available, the theoretical chromatographic retention time as well as the standard fragmentation pathway are also compared and an undoubtedly identification can be done. In contrast, authors who work with simple HRMS analyzers, such as TOF, normally resort to LRMS analyzers able to perform tandem experiments such as QqQ or IT instruments in order to obtain this fragmentation pathway data [20].

Data treatment selection is strongly related to the food analytical challenge that has to be solved. In some applications, such as some adulteration food frauds, particular biomarkers are significantly discriminant between native and adulterated food samples and their determination allows the detection of that illegal practice. As an example, characteristic and endogenous glycosylated kaempferol derivatives were used as authenticity markers able to detect and quantify the adulteration in saffron samples regardless the substance used as adulterant [15]. Depending on the compound selected as biomarker, a limit of detection for the adulteration content between 0.2 and 2.5% was achieved. However, a statistical procedure like ANOVA is usually needed for the evaluation of the significance of difference for targeted compounds among different types of samples. Guo et al. [16] used this strategy to verify the capability of targeted polyphenols to classify kiwifruit juices according to their variety and geographical origin. Even though the presence of certain polyphenols was significantly different in each case, none of them was able to cluster the samples by itself. It is in these types of situations when multivariate data analysis techniques have gained relevance, as they can combine information regarding the content of a large number of compounds. As an example, an unsupervised PCA was applied in order to study the geographical and botanical origin of Serbian red spice paprika samples (Lemeška and Lakošnička varieties) by using the concentration of 25 polyphenols, obtained by a targeted LC-HRMS method, and 13 carbohydrates, quantified by high performance anion exchange chromatography with pulsed amperometric detection [18]. The scores plots for the first two PCs, which explain 52.75% of the total variance, show a good discrimination between samples of different origin. In fact, loadings plot revealed a strong influence of particular phenolic acids and flavonoid glycosides in the separation of Lemeška paprika samples, while flavonoid aglycones and carbohydrates mainly affected the discrimination of the Lakošnička variety. In addition, Lakošnička samples were also classified according to their harvesting year (2012 or 2013). On the other hand, Barbosa et al. [19] also used chemometric multivariate analysis tools, but with the aim of preventing the possible adulteration of cranberry-based commercial pharmaceuticals with other ineffective and less expensive fruit-based extracts. Cranberry (*Vaccinium macrocarpon*) and its derivatives are known to prevent urinary tract infections as they contain A-type proanthocyanidins, which exhibit bioactive activity; in contrast, less expensive fruits like blueberry, raspberry, or grapes are richer in B-type proanthocyanidins, which does not exhibit this beneficial effect in human health. The authors developed a targeted polyphenolic UHPLC-HRMS (Orbitrap) method for the classification, authentication and detection of frauds in fruit-based extracts, and a total of 106 samples including cranberry-, grape-, blueberry-, and raspberry-based natural products, as well as cranberry-based pharmaceutical preparations presented in different formats were analyzed. Then, a built user-accurate mass database of 53 polyphenols containing spectral data and several confirmation parameters (accurate mass measurements, isotopic pattern matches, product-ion scan spectra, and chromatographic retention time) was applied for screening and

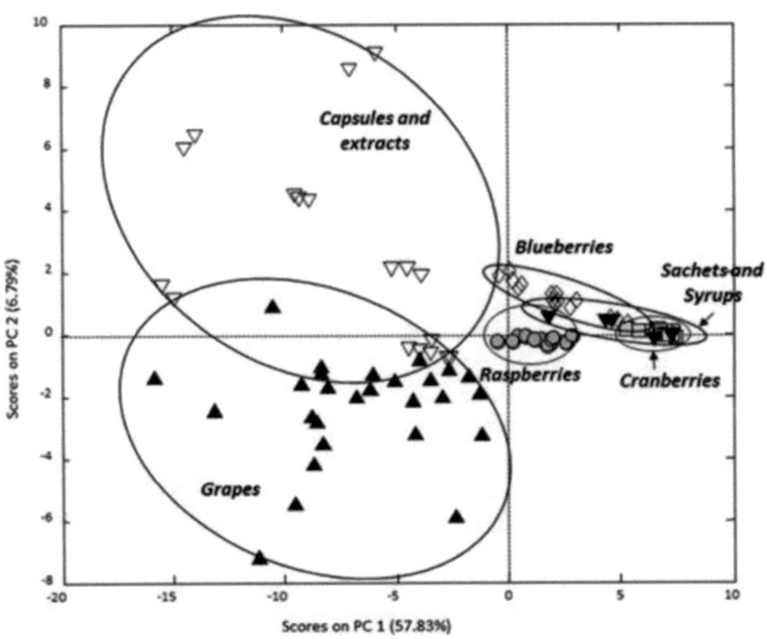

Figure 2.
PCA score plot of PC1 vs. PC2 obtained using UHPLC-HRMS (Orbitrap) polyphenolic profiles for the classification and authentication of fruit-based extracts and cranberry-based pharmaceutical preparations. Reproduced with permission from Ref. [19]. Copyright (2018) American Chemical Society.

confirmation purposes. The obtained polyphenolic content, which was described in a data matrix containing the peak area of each detected target compound, was employed as chemical descriptors by PCA. As can be seen in **Figure 2**, the obtained scores plots show a good discrimination between cranberry-based samples in front of any other fruit-based sample, showing the ability of the developed method to clearly authenticate the fruit extracts according to the type of fruit employed.

Moreover, a PLS model was developed to predict and quantify fraud levels of adulterant fruit (grape, blueberry, and raspberry) extracts in cranberry-fruit extracts, reaching calibration errors below 0.01% and prediction errors in the range of 2.71 and 5.96%, demonstrating the suitability of polyphenolic targeted LC-HRMS methods in food integrity and authenticity.

3. Non-targeted LC-HRMS (metabolomics) approaches

Modern HRMS analyzers, such as TOF and Orbitrap (and their hybrid configurations), have focused the analysis of food samples from a totally different perspective as did before, mainly due to their high capacity to generate and register a large amount of information, especially when working in full-scan or in data-dependent scan acquisition modes. In fact, in the last years, there has been a trend toward non-targeted LC-HRMS metabolomic approaches, either by studying the metabolomic profiles of food, where chemometrics plays an essential role in the data treatment, or by a retrospective analysis in order to identify unknown compounds that could become new food biomarkers. Although non-targeted metabolomic approaches are potentially much more informative than targeted approaches in practice, the annotation of the features either obtained by using databases or by matching with pure standard data is frequently required. However, in many cases, metabolomic fingerprinting is enough to classify and discriminate among food samples; therefore, further metabolite identification is not needed. Moreover, it should be taken into account that when dealing with non-targeted metabolomics, the final annotated metabolites are strongly dependent on the global experimental approached employed (including sample treatment, separation and detection, as well as the specific instrumentation used). As an example, Díaz et al. [22] studied the influence of the global approach on the final annotated metabolites in non-targeted metabolomic analysis of 42 red wine samples (from three different Spanish PDO) when comparing two LC-MS interplatforms that differed in columns, mobile phases, gradients, chromatographs, mass spectrometers (Q-Orbitrap [Platform #1] and Q-TOF [Platform #2]), data processing, and marker selection protocols. **Figure 3** shows a scheme of the experimental workflow described by the authors. The authors showed that despite the ability of the platforms to distinguish the wine classes at both the spectral and the annotated metabolite level, a strong divergence among the annotated metabolites involved in the discrimination was found. For example, at the annotated features level, PDO classes were separated using both experimental setups (**Figure 4a** and **b**). When annotated metabolite level was employed, a total of 9 and 8 features were identified for Platforms #1 and #2, respectively, although none of them was common. PCA models built using only these annotated features resulted in a clear separation when the Q-TOF was employed (**Figure 4d**), but with the Q-Orbitrap, wines from Ribera del Duero PDO and Rioja PDO were not completely separated (**Figure 4c**). When results obtained using both compared platforms were considered, the resulting PCA model performed including only the annotated features common for both platforms showed a high degree of similarity between them (**Figure 4e** and **f**).

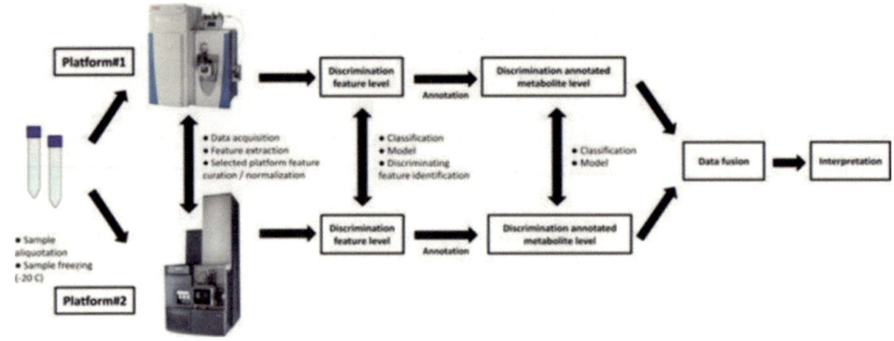

Figure 3.
Summary of the experimental workflow compared. Reproduced with permission from Ref. [22]. Copyright (2016) Elsevier.

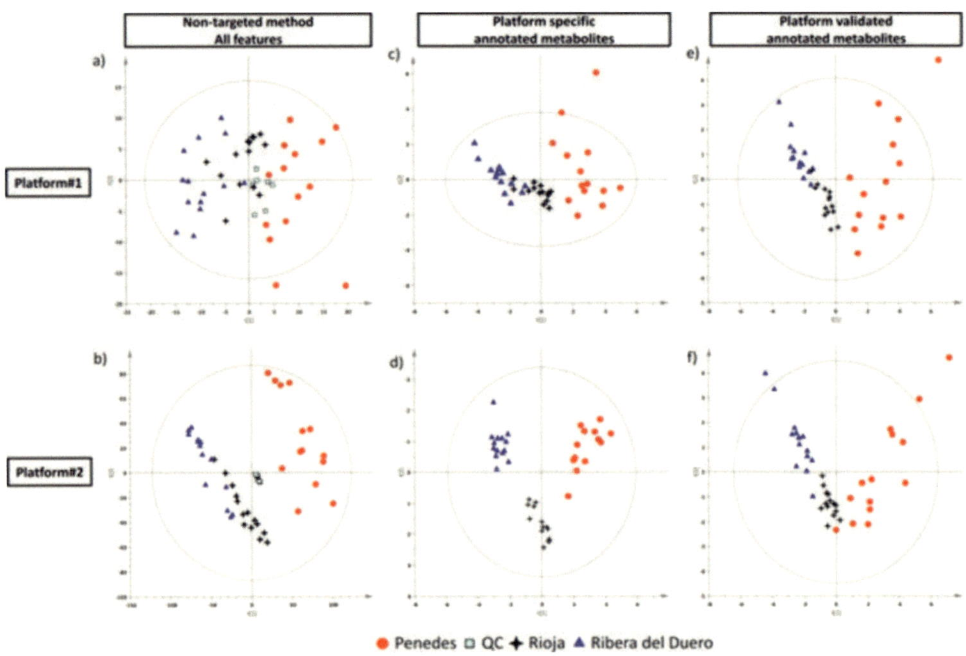

Figure 4.
Plots of scores obtained for the PCA analysis for Platforms #1 and #2 at the all features level (a and b), platform specific annotated metabolite level (c and d) and interplatform validated metabolites (e and f). Reproduced with permission from Ref. [22]. Copyright (2016) Elsevier.

This study shows the complications that may arise on the comparison of non-targeted metabolomic platforms even when metabolite focused approaches are used in the identification.

Table 3 summarizes a wide variety on non-targeted LC-HRMS applications in food integrity and authenticity. As previously indicated, these non-targeted methodologies can be considered as blind approaches toward the unknown metabolomic composition of a particular group of samples. For that reason, the selection and optimization of the chromatographic separation or ionization technique have to be done conscientiously as they will delimit the detected compounds according to their hydrophobicity and ionization capacity. As can be seen in the table, reversed-phase stationary columns are also usually chosen to conduct the chromatographic separation in non-targeted LC-HRMS approaches. This is because this separation mode

Sample	Chromatographic separation and mass spectrometry	Data analysis	Ref.
Oregano	Acquity HSS T3 column (100 × 2.1 mm, 1.8 µm) Gradient elution (0.4 mL·min^{-1}): (A) water with 0.1% formic acid (B) methanol with 0.1% formic acid H-ESI (±) Q-TOF (full-scan mode 50–1200 *m/z*)	PCA and OPLS-DA	[39]
Lamb	ZIC p-HILIC column (100 × 2.1 mm, 5 µm) Gradient elution (0.25 mL·min^{-1}): (A) acetonitrile with 0.1% formic acid (B) water with 16mM ammonium formate ESI (±) Orbitrap (full-scan mode 55–1100 *m/z*)	Fold change, t-test and PCA	[44]
Tomato juices	C30 column (250 × 4.6 mm, 3 µm) Gradient elution (1.3 mL·min^{-1}): (A) methanol:methyl tert-butyl ether:water with 2% ammonium acetate 60:35:5 (v/v/v) (B) methyl tert-butyl ether:methanol:water with 2% ammonium acetate (v/v/v) APCI (+) Q-TOF (full-scan mode 100–1700 *m/z*)	Fold change and t-test	[41]
Coffee	Ascentis Express C18 column (150 × 2.1 mm, 2.7 µm) Gradient elution (0.2 mL·min^{-1}): (A) water with 0.1% formic acid (B) acetonitrile with 0.1% formic acid H-ESI (±) Q-TOF (full-scan mode 100–1700 *m/z*)	PCA and PLS-DA	[23]
Tomato	Phenomenex Luna C8 column (100 × 2 mm, 3 µm) Gradient elution (0.35 mL·min^{-1}): (A) water:methanol 98:2 (v/v) (B) methanol:water 98:2 (v/v) H-ESI (+) Q-Orbitrap (full-scan mode 74–1100 *m/z*)	PCA	[40]
Beef meat	Hypersil Gold C18 column (100 × 2.1 mm, 1.9 µm) Gradient elution (0.4 mL·min^{-1}): (A) water (B) methanol ESI (+) LTQ-Orbitrap (full-scan mode 50–2000 *m/z*)	PCA and PLS-DA	[24]
Citrus fruit/ fruit juices	Acquity UPLC BEH C18 column (100 × 2.1 mm, 1.7 µm) Gradient elution (0.4 mL·min^{-1}): (A) water with 10 mM ammonium acetate (B) acetonitrile H-ESI (+) Q-TOF (full-scan mode 50–1200 *m/z*)	PCA, PLS-DA and SIMCA	[31]
Wild strawberry	XSelect CSH C18 column (150 × 2.1 mm, 1.6 µm) Gradient elution (0.2 mL·min^{-1}): (A) water with 0.1% formic acid (B) acetonitrile:water 80:20 (v/v) with 0.1% formic acid H-ESI (±) LTQ-Orbitrap (full-scan mode 200–1600 *m/z*)	PCA	[32]
Eggs	Phenomenex Luna Omega C18 column (150 × 2.1 mm, 3.5 µm) Gradient elution (0.3 mL·min^{-1}): (A) water with 0.1% formic acid and 5 mM ammonium formate (B) methanol with 0.1% formic acid and 5 mM ammonium formate H-ESI (±) Q-Orbitrap (full-scan mode 75–1000 *m/z*)	PCA	[33]
Garlic	Mediterranea Sea C18 column (150 × 4.6 mm, 5 µm) Gradient elution (0.7 mL·min^{-1}): (A) water with 0.1% formic acid (B) methanol with 0.1% formic acid H-ESI (±) Q-TOF (full-scan mode 30–1700 *m/z*)	ANOVA and PCA	[34]

Sample	Chromatographic separation and mass spectrometry	Data analysis	Ref.
Parmigiano Reggiano cheese	Kinetex XB C18 column (100 × 3 mm, 2.6 μm) Gradient elution (0.5 mL·min^{-1}): (A) water with 0.2% formic acid (B) acetonitrile with 0.2% formic acid H-ESI (−) Orbitrap (full-scan mode 50–900 *m/z*)	PCA Class Method	[35]
Cheeses	Hypersil Gold C18 column (100 × 2.1 mm, 1.9 μm) Gradient elution (0.4 mL·min^{-1}): (A) water with 0.1% acetic acid (B) acetonitrile with 0.1% acetic acid H-ESI (±) Orbitrap (full-scan mode 65–1000 *m/z*)	PCA	[36]
Myrtle berry	XSelect CSH C18 column (150 × 2.1 mm, 3.5 μm) Gradient elution (0.2 mL·min^{-1}): (A) water with 0.1% formic acid (B) acetonitrile with 0.1% formic acid ESI (−) LTQ-Orbitrap (full-scan mode 200–1600 *m/z*)	PCA	[37]
Saffron	Ascentis Express C18 column (100 × 2.1 mm, 2.7 μm) Gradient elution (0.4 mL·min^{-1}): (A) water with 0.1% formic acid or 10 mM ammonium formate (B) acetonitrile with 0.1% formic acid or 10 mM ammonium formate H-ESI (±) Q-TOF (full-scan mode 100–1700 *m/z*)	PCA, PLS-DA and OPLS-DA	[38]
Eggs	Thermo Scientific Accucore C18 column (100 × 2.1 mm, 2.6 μm) Gradient elution (0.3 mL·min^{-1}): (A) water with 0.1% formic acid and 5 mM ammonium acetate (B) methanol with 0.1% formic acid and 5 mM ammonium acetate H-ESI (+) Q-TOF (full-scan mode 100–1000 *m/z*)	PCA	[25]
Olive oil	Acclaim RSLC C18 column (100 × 2.1 mm, 2.2 μm) Gradient elution (0.2 mL·min^{-1}): (A) water:methanol 90:10 (v/v) with 5 mM ammonium acetate (B) methanol with 5 mM ammonium acetate H-ESI (−) Q-TOF (full-scan mode 50–1000 *m/z*)	PCA	[26]
Saffron	Phenomenex Kinetex C18 column (100 × 2.1 mm, 1.7 μm) Gradient elution (0.4–0.6 mL·min^{-1}): (A) water with 5 mM ammonium formate or acetate (B) methanol H-ESI (±) Q-TOF (full-scan mode 100–1200 *m/z*)	PCA and OPLS-DA	[27]
Honey	Hypersil Gold C18 column (100 × 2.1 mm, 1.9 μm) Gradient elution (0.3 mL·min^{-1}): (A) water with 0.1% formic acid (B) acetonitrile with 0.1% formic acid H-ESI (±) Q-Orbitrap (full-scan mode 80–1200 *m/z*)	PCA and PLS-DA	[28]
Beer	Hypersil Gold aQ column (100 × 2.1 mm, 1.9 μm) Gradient elution (0.6 mL·min^{-1}): (A) acetonitrile with 0.1% formic acid (B) water with 0.1% formic acid H-ESI (±) LTQ-Orbitrap (full-scan mode 50–1000 *m/z*)	PCA and PLS-DA	[43]
Tiger nut	BEH C18 column (100 × 2.1 mm, 1.7 μm) Gradient elution (0.4–0.5 mL·min^{-1}): (A) water:methanol 95:5 (v/v) with 0.1% formic acid and 5 mM ammonium formate (B) 2-propanol:methanol:water 65:30:5 (v/v/v) with 0.1% formic acid and 5 mM ammonium formate H-ESI (±) Q-TOF (full-scan mode 100–1200 *m/z*)	PCA and OPLS-DA	[29]

Sample	Chromatographic separation and mass spectrometry	Data analysis	Ref.
Tea	Hypersil Gold C18 column (100 × 2.1 mm, 1.9 μm) Gradient elution (0.2 mL·min^{-1}): (A) water with 0.1% formic acid (B) acetonitrile H-ESI (±) Q-Orbitrap (full-scan mode 100–1500 *m/z*)	PCA, HCA and PLS-DA	[30]
Chicken	Hypersil Gold aQ column (100 × 2.1 mm, 1.9 μm) Gradient elution (0.3 mL·min^{-1}): (A) water with 0.1% formic acid (B) acetonitrile with 0.1% formic acid H-ESI (±) Q-TOF (full-scan mode 100–1000 *m/z*)	PCA	[42]

Soft independent modeling by class analogy (SIMCA).

Table 3.
Recent advances of non-targeted LC-HRMS methodologies in food integrity and authenticity.

provides a great chromatographic separation of semi-polar metabolites, which comprise a wide number of compounds (phenolic acids, flavonoids, alkaloids, or glycosylated species) that have proved to be useful and interesting for the authentication of food samples. Although C18 is generally proposed [23–38], some strategies have employed other reversed-phases such as HSS T3 [39], C8 [40], and C30 [41]. For instance, Black et al. [39] used a HSS T3 column for the separation of potential biomarkers able to identify adulteration in oregano samples, while Martinez et al. [40] proposed a C8 column with the aim to separate unknown food markers for the discrimination of tomato samples obtained in organic or conventional crops.

On the other hand, the enhanced bioavailability of *tangerine* tomato lycopene in front of red tomato lycopene, which is attributed in part to tetra-cis lycopene geometric configuration in *tangerine* variety rather than all-trans configuration of red variety, took Cichon et al. [41] to study and compare the metabolomic phytochemical composition between them. A C30 column was chosen for the chromatographic separation taking advantage of its high selectivity toward the separation of hydrophobic structurally related isomeric compounds. Even though RPLC is widely exploited in these applications, there is a range of polar metabolites (amino acids, carbohydrates, sugars, amines, or organic acids) that normally elute in the solvent front. Thus, polar endcapped C18 [42, 43] or HILIC [44] columns can be employed for their study as they offer an alternative selectivity. As an example, Gallart-Ayala et al. [43] used a polar-endcapped C18 column for the separation of moderate polar compounds in order to compare beers obtained by different brewing procedures.

Accordingly to targeted approaches, H-ESI is also the most employed ionization technique in the non-targeted LC-HRMS approach. However, in this case, normally both positive and negative ionization modes are studied [23, 27–30, 32–34, 36, 38, 39, 42–44], since it is not intended to find determined compounds but rather to study which of them provides a solution to the food integrity and authentication challenge, even if they are not identified. In some applications, other API sources can offer an interesting ionization range as well as less matrix effect. For instance, instead of H-ESI, APCI operated in positive ionization mode was used to analyze lipophilic extracts of tomato juices detecting a total of 423 compounds among which 352 were significantly different between the two types of juices studied [41].

As described in **Table 3**, hybrid HRMS analyzers (Q-Orbitrap, LTQ-Orbitrap and Q-TOF) are also widely employed in non-targeted LC-HRMS methodologies. The possibility to study the fragmentation of unknown molecular features allows their identification and confirmation in order to establish them as future targeted compounds for particular applications. Even though single HRMS analyzers do not provide fragmentation data of the detected ions, metabolomics data can solve authenticity problematics without the identification of any compound (fingerprinting strategy) as previously commented [35].

The first step of non-targeted LC-HRMS approaches data treatment is the conversion of raw data in a matrix built by retention time, *m/z* values and the area or signal of each peak detected. Sometimes, chemical interferences are removed from the matrix by fixing some parameters to be achieved such as mass tolerance for peak alignment, total intensity threshold, maximum peak shift, and S/N threshold. At this point, the generated matrix can be treated by univariate or multivariate data analysis. For instance, d'Urso et al. [32], who aimed to compare wild strawberry samples of different geographical origin (Sarno and Petina, Italy), growing conditions (spontaneous and cultivated populations), and germplasm (autochthonous and non-autochthonous), created a unique data matrix from raw data obtained in both positive and negative ionization mode in the performed LC-HRMS analysis following a data fusion procedure. PCA was then applied to the data matrix obtaining a scores plot that clearly discriminates between spontaneous and cultivated samples regardless the other variables. Moreover, a good classification was also observed for the five groups of samples studied, which were different combinations of the above geographical origin, growing conditions and germplasm mentioned variables.

Anyways, when the objective of the study is the identification of molecular features that could behave as a biomarker in food integrity and authenticity, the matrix needs to be reduced. Thus, measures like the elimination of those molecular features that are not detected in a minimum percentage of the samples or of those that are not observed in the quality controls, which usually consists in a mix formed by a constant volume of all the analyzed samples, are normally implemented in the non-targeted LC-HRMS workflow.

As an example, Cavanna et al. [33], whose objective was the identification and selection of biomarkers responsible of the freshness of egg products, proposed a first reduction of data matrix by establishing some critical parameters values: (i) precursor ion deviation of 5 and 10 ppm for negative and positive runs, respectively, (ii) maximum peak shift of 0.3 min, (iii) a total intensity threshold of 1,000,000 AU, and (iv) a 30% of relative intensity tolerance used for isotope search. The authors removed the molecular features that showed a coefficient of variation bigger than 40% in the quality control sample, which was prepared by mixing 10 μL of each extract sample and was injected at the beginning of the sequence as well as every 10 samples analyzed. As a clear separation between fresh and non-fresh egg samples was observed when making a PCA study on positive and negative ionization modes with the reduced matrixes, the authors then applied supervised OPLS-DA. As can be seen in **Figure 5**, an expected increase in the discrimination was achieved.

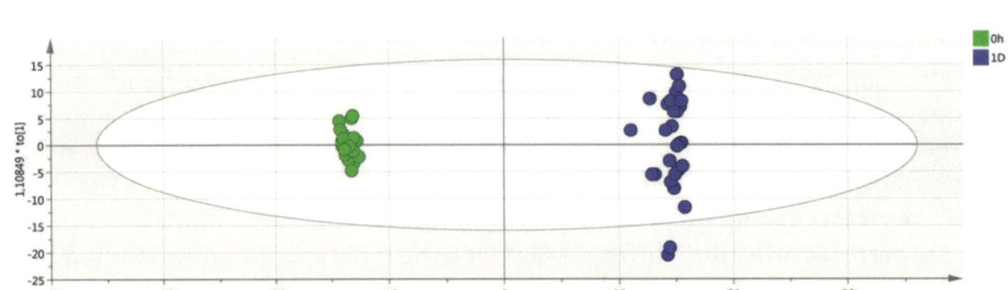

Figure 5.
ESI + OPLS-DA scores plot of the fresh samples against the "1 day" samples. Left area dots (0 h), fresh samples; right area dots (1D), "1 day" samples. Reproduced from Ref. [33]. Open Access Journals.

S-plots, which correspond to OPLS-DA loading plots, and variable importance in projection (VIP) values were used to select the most significant features in the clusterization of the samples. In order to identify those molecular features, exact mass, the isotopic pattern and MS/MS fragmentation were studied. As a result, 12 compounds were completely identified (standard injection confirm their identity) and 19 were tentatively identified by the authors.

4. Summary and concluding remarks

The role of LC-MS and LC-HRMS methodologies to address food integrity and authenticity have been presented and discussed by means of some selected applications published in the last years.

Most of the methods described in the literature opt for RPLC with mainly C18 columns, with gradient elution using acidified aqueous solutions and methanol or acetonitrile as mobile phase components, probably due to the strong capacity of this separation mode when dealing with low molecular weight chemicals with a relatively wide range of polarities. The use of other stationary phases such as C18 amide or perfluorinated columns are also proposed in some specific applications.

ESI continues to be the ionization source of choice when dealing with LC-MS and LC-HRMS analysis of food products, although in some cases other API sources are also employed. APPI has shown to provide similar or slightly better sensitivity for some specific applications, such as in the case of the determination of polyphenols, but it resulted in a very feasible option when addressing the characterization and classification of natural extracts due to the higher robustness of APPI source in the presence of matrix effect. Therefore, although it has not been widely exploited in food integrity and authenticity issues up to now, it is strongly recommended because of the sample matrix complexity of foodstuffs.

Regarding the mass analyzers, QqQ and IT instruments are the chosen ones when LRMS is employed, and TOF and Orbitrap analyzers for HRMS applications. However, the selection of LC-MS or LC-HRMS methods usually depends on the targeted or non-targeted approach. When targeted strategies are proposed, some specific biochemical food components are determined as food features to address food integrity and authenticity, requiring a quantitation step using standards for each targeted component. In those cases, LC-MS(/MS) methodologies, mainly using QqQ instruments, are very appropriate due to the low sensitivity attainable with these analyzers, and their good performance for quantitative analysis. Obviously, LC-HRMS methods providing higher resolution and accurate mass measurements are also a very good option for targeted food analysis, although it is more expensive and requires a more specialized staff. In order to achieve sample characterization and authentication, the comparison of the content and distribution of the targeted chemicals is sometimes enough, but the use of chemometric methods to try to find food feature similarities between the analyzed samples is highly recommended, especially when both the number of samples and the number of targeted bioactive substances increase.

In many applications, the quantitation of some chemicals may be a difficult task due to food matrix complexity, especially due to the possibility of unknown interfering compounds. In those cases, non-targeted approaches (based on metabolomic fingerprinting) using LC-HRMS have shown to be the best option to address food integrity and authenticity. As non-targeted analysis is performed, the high resolution and accurate mass measurements attainable with TOF and Orbitrap instruments are required. In non-targeted approaches, the measurement of peak intensity values as a function of m/z and retention times is frequently enough to achieved food integrity and authenticity. Obviously, due to the huge amount of data obtained, especially when working in full-scan mode, the use of chemometrics is mandatory. Nevertheless, it has been reported that when dealing with metabolomic HRMS methodologies, the final annotated metabolites are strongly dependent on the global experimental approach employed (sample treatment, separation and detection, instrumentation employed, etc.). This is very important when searching for possible food biomarkers, as those will depend on the methodology used.

In conclusion, targeted and non-targeted LC-MS and LC-HRMS methodologies, especially in combination with multivariate chemometric methods, are powerful tools to address a hot topic nowadays such as food integrity and authenticity, and the number of publications in this field will continue to increase in the near future.

Acknowledgements

This work was supported by the Agency for the Administration of University and Research Grants (Generalitat de Catalunya, Spain) under the Project 2017 SGR-310.

Author details

Guillem Campmajó[1], Nerea Núñez[1] and Oscar Núñez[1,2,3]*

1 Department of Chemical Engineering and Analytical Chemistry, University of Barcelona, Barcelona, Spain

2 Research Institute in Food Nutrition and Food Safety, University of Barcelona, Barcelona, Spain

3 Serra Húnter Program, Generalitat de, Catalunya

*Address all correspondence to: oscar.nunez@ub.edu

References

[1] Moore JC, Spink J, Lipp M. Development and application of a database of food ingredient fraud and economically motivated adulteration from 1980 to 2010. Journal of Food Science. 2012;**77**(4):R118-R126. DOI: 10.1111/j.1750-3841.2012.02657.x

[2] Núñez O, Gallart-Ayala H, Martins CPB, Lucci P, editors. Fast Liquid Chromatography-Mass Spectrometry Methods in Food and Environmental Analysis. London, UK: Imperial College Press; 2015

[3] Saurina J, Sentellas S. Determination of phenolic compounds in food matrices: Application to characterization and authentication. In: Núñez O, Gallart-Ayala H, Martins CPB, Lucci P, editors. Fast Liquid Chromatography-Mass Spectrometry Methods in Food and Environmental Analysis. London, UK: Imperial College Press; 2015. pp. 517-547

[4] Brown SD, Tauler R, Walczak B. Comprehensive Chemometrics. Chemical and Biochemical Data Analysis. Vol. 3. Amsterdam, The Netherlands: Elsevier; 2009

[5] Lucci P, Saurina J, Núñez O. Trends in LC-MS and LC-HRMS analysis and characterization of polyphenols in food. TrAC Trends in Analytical Chemistry. 2017;**88**:1-24. DOI: 10.1016/j.trac.2016.12.006

[6] Brighenti V, Groothuis SF, Prencipe FP, Amir R, Benvenuti S, Pellati F. Metabolite fingerprinting of *Punica granatum* L. (pomegranate) polyphenols by means of high-performance liquid chromatography with diode array and electrospray ionization-mass spectrometry detection. Journal of Chromatography. A. 2017;**1480**:20-31. DOI: 10.1016/j.chroma.2016.12.017

[7] Seraglio SKT, Valese AC, Daguer H, Bergamo G, Azevedo MS, Gonzaga LV, et al. Development and validation of a LC-ESI-MS/MS method for the determination of phenolic compounds in honeydew honeys with the diluted-and-shoot approach. Food Research International. 2016;**87**:60-67. DOI: 10.1016/j.foodres.2016.06.019

[8] Alarcón-Flores MI, Romero-González R, Martínez Vidal JL, Garrido Frenich A. Determination of phenolic compounds in artichoke, garlic and spinach by ultra-high-performance liquid chromatography coupled to tandem mass spectrometry. Food Analytical Methods. 2014;**7**(10):2095-2106. DOI: 10.1007/s12161-014-9852-4

[9] Becerra-Herrera M, Lazzoi MR, Sayago A, Beltrán R, Del Sole R, Vasapollo G. Extraction and determination of phenolic compounds in the berries of Sorbus americana marsh and Lonicera oblongifolia (Goldie) hook. Food Analytical Methods. 2015;**8**(10):2554-2559. DOI: 10.1007/s12161-015-0151-5

[10] Ribas-Agustí A, Cáceres R, Gratacós-Cubarsí M, Sárraga C, Castellari M. A validated HPLC-DAD method for routine determination of ten phenolic compounds in tomato fruits. Food Analytical Methods. 2012;**5**(5):1137-1144. DOI: 10.1007/s12161-011-9355-5

[11] Puigventós L, Navarro M, Alechaga É, Núñez O, Saurina J, Hernández-Cassou S, et al. Determination of polyphenolic profiles by liquid chromatography-electrospray-tandem mass spectrometry for the authentication of fruit extracts. Analytical and Bioanalytical Chemistry. 2015;**407**(2):597-608. DOI: 10.1007/s00216-014-8298-2

[12] Parets L, Alechaga É, Núñez O, Saurina J, Hernández-Cassou S, Puignou L. Ultrahigh pressure liquid chromatography-atmospheric

pressure photoionization-tandem mass spectrometry for the determination of polyphenolic profiles in the characterization and classification of cranberry-based pharmaceutical preparations and natural extracts. Analytical Methods. 2016;**8**(22): 4363-4378. DOI: 10.1039/C6AY00929H

[13] Alakolanga AGAW, Siriwardene AMDA, Savitri Kumar N, Jayasinghe L, Jaiswal R, Kuhnert N. LC-MSn identification and characterization of the phenolic compounds from the fruits of *Flacourtia indica* (Burm. F.) Merr. and *Flacourtia inermis* Roxb. Food Research International. 2014;**62**:388-396. DOI: 10.1016/j.foodres.2014.03.036

[14] Shanmugam S, Gomes IA, Denadai M, Lima BDS, Araújo AADS, Narain N, et al. UHPLC-QqQ-MS/MS identification, quantification of polyphenols from Passiflora subpeltata fruit pulp and determination of nutritional, antioxidant, α-amylase and α-glucosidase key enzymes inhibition properties. Food Research International. 2018;**108**:611-620. DOI: 10.1016/j.foodres.2018.04.006

[15] Guijarro-Díez M, Castro-Puyana M, Crego AL, Marina ML. A novel method for the quality control of saffron through the simultaneous analysis of authenticity and adulteration markers by liquid chromatography-(quadrupole-time of flight)-mass spectrometry. Food Chemistry. 2017;**228**:403-410. DOI: 10.1016/j.foodchem.2017.02.015

[16] Guo J, Yuan Y, Dou P, Yue T. Multivariate statistical analysis of the polyphenolic constituents in kiwifruit juices to trace fruit varieties and geographical origins. Food Chemistry. 2017;**232**:552-559. DOI: 10.1016/j.foodchem.2017.04.037

[17] Zhang J, Yu Q, Cheng H, Ge Y, Liu H, Ye X, et al. Metabolomic approach for the authentication of berry fruit juice by liquid chromatography quadrupole time-of-flight mass spectrometry coupled to chemometrics. Journal of Agricultural and Food Chemistry. 2018;**66**(30):8199-8208. DOI: 10.1021/acs.jafc.8b01682

[18] Mudrić S, Gašić UM, Dramićanin AM, Ćirić I, Milojković-Opsenica DM, Popović-Đorđević JB, et al. The polyphenolics and carbohydrates as indicators of botanical and geographical origin of Serbian autochthonous clones of red spice paprika. Food Chemistry. 2017;**217**:705-715. DOI: 10.1016/j.foodchem.2016.09.038

[19] Barbosa S, Pardo-Mates N, Hidalgo-Serrano M, Saurina J, Puignou L, Núnez O. Detection and quantitation of frauds in the authentication of cranberry-based extracts by UHPLC-HRMS (Orbitrap) polyphenolic profiling and multivariate calibration methods. Journal of Agricultural and Food Chemistry. 2018;**66**(35):9353-9365. DOI: 10.1021/acs.jafc.8b02855

[20] Megías-Pérez R, Grimbs S, D'Souza RN, Bernaert H, Kuhnert N. Profiling, quantification and classification of cocoa beans based on chemometric analysis of carbohydrates using hydrophilic interaction liquid chromatography coupled to mass spectrometry. Food Chemistry. 2018;**258**:284-294. DOI: 10.1016/j.foodchem.2018.03.026

[21] Zhu D, Nyström L. Differentiation of rice varieties using small bioactive lipids as markers. European Journal of Lipid Science and Technology. 2015;**117**(10):1578-1588. DOI: 10.1002/ejlt.201500089

[22] Díaz R, Gallart-Ayala H, Sancho JV, Núñez O, Zamora T, Martins CPB, et al. Told through the wine: A liquid chromatography-mass spectrometry interplatform comparison reveals the influence of the global approach on the final annotated metabolites in non-targeted metabolomics. Journal of

Chromatography. A. 2016;**1433**:90-97. DOI: 10.1016/j.chroma.2016.01.010

[23] Pérez-Míguez R, Sánchez-López E, Plaza M, Castro-Puyana M, Marina ML. A non-targeted metabolomic approach based on reversed-phase liquid chromatography—Mass spectrometry to evaluate coffee roasting process. Analytical and Bioanalytical Chemistry. 2018;**410**(30):7859-7870. DOI: 10.1007/s00216-018-1405-z

[24] Trivedi DK, Hollywood KA, Rattray NJW, Ward H, Trivedi DK, Greenwood J, et al. Meat, the metabolites: An integrated metabolite profiling and lipidomics approach for the detection of the adulteration of beef with pork. The Analyst. 2016;**141**(7):2155-2164. DOI: 10.1039/C6AN00108D

[25] Johnson AE, Sidwick KL, Pirgozliev VR, Edge A, Thompson DF. Metabonomic profiling of chicken eggs during storage using high-performance liquid chromatography—Quadrupole time-of-flight mass spectrometry. Analytical Chemistry. 2018;**90**(12):7489-7494. DOI: 10.1021/acs.analchem.8b01031

[26] Kalogiouri NP, Aalizadeh R, Thomaidis NS. Application of an advanced and wide scope non-target screening workflow with LC-ESI-QTOF-MS and chemometrics for the classification of the Greek olive oil varieties. Food Chemistry. 2018;**256**:53-61. DOI: 10.1016/j.foodchem.2018.02.101

[27] Rubert J, Lacina O, Zachariasova M, Hajslova J. Saffron authentication based on liquid chromatography high resolution tandem mass spectrometry and multivariate data analysis. Food Chemistry. 2016;**204**:201-209. DOI: 10.1016/j.foodchem.2016.01.003

[28] Li Y, Jin Y, Yang S, Zhang W, Zhang J, Zhao W, et al. Strategy for comparative untargeted metabolomics reveals honey markers of different floral and geographic origins using ultrahigh-performance liquid chromatography-hybrid quadrupole-orbitrap mass spectrometry. Journal of Chromatography. A. 2017;**1499**:78-89. DOI: 10.1016/j.chroma.2017.03.071

[29] Rubert J, Hurkova K, Stranska M, Hajslova J. Untargeted metabolomics reveals links between Tiger nut (*Cyperus esculentus* L.) and its geographical origin by metabolome changes associated with membrane lipids. Food Additives and Contaminants—Part A Chemistry, Analysis, Control, Exposure and Risk Assessment. 2018;**35**(4):605-613. DOI: 10.1080/19440049.2017.1400694

[30] Xin Z, Ma S, Ren D, Liu W, Han B, Zhang Y, et al. UPLC-Orbitrap-MS/MS combined with chemometrics establishes variations in chemical components in green tea from Yunnan and Hunan origins. Food Chemistry. 2018;**266**:534-544. DOI: 10.1016/j.foodchem.2018.06.056

[31] Jandrić Z, Cannavan A. An investigative study on differentiation of citrus fruit/fruit juices by UPLC-QToF MS and chemometrics. Food Control. 2017;**72**:173-180. DOI: 10.1016/j.foodcont.2015.12.031

[32] D'Urso G, Maldini M, Pintore G, d'Aquino L, Montoro P, Pizza C. Characterisation of *Fragaria vesca* fruit from Italy following a metabolomics approach through integrated mass spectrometry techniques. LWT—Food Science and Technology. 2016;**74**:387-395. DOI: 10.1016/j.lwt.2016.07.061

[33] Cavanna D, Catellani D, Dall'Asta C, Suman M. Egg product freshness evaluation: A metabolomic approach. Journal of Mass Spectrometry. 2018;**53**(9):849-861. DOI: 10.1002/jms.4256

[34] Molina-Calle M, Sanchez de Medina V, Priego-Capote F, Luque de Castro MD. Establishing compositional differences between fresh and black garlic by a metabolomics approach based on LC-QTOF MS/MS analysis. Journal of Food Composition and Analysis. 2017;**62**:155-163. DOI: 10.1016/j.jfca.2017.05.004

[35] Popping B, De Dominicis E, Dante M, Nocetti M. Identification of the geographic origin of Parmigiano Reggiano (P.D.O.) cheeses deploying non-targeted mass spectrometry and chemometrics. Food. 2017;**6**(2):13. DOI: 10.3390/foods6020013

[36] Le Boucher C, Courant F, Royer AL, Jeanson S, Lortal S, Dervilly-Pinel G, et al. LC-HRMS fingerprinting as an efficient approach to highlight fine differences in cheese metabolome during ripening. Metabolomics. 2015;**11**(5):1117-1130. DOI: 10.1007/s11306-014-0769-0

[37] D'Urso G, Sarais G, Lai C, Pizza C, Montoro P. LC-MS based metabolomics study of different parts of myrtle berry from Sardinia (Italy). Journal of Berry Research. 2017;7(3):217-229. DOI: 10.3233/JBR-170158

[38] Guijarro-Díez M, Nozal L, Marina ML, Crego AL. Metabolomic fingerprinting of saffron by LC/MS: Novel authenticity markers. Analytical and Bioanalytical Chemistry. 2015;**407**(23):7197-7213. DOI: 10.1007/s00216-015-8882-0

[39] Black C, Haughey SA, Chevallier OP, Galvin-King P, Elliott CT. A comprehensive strategy to detect the fraudulent adulteration of herbs: The oregano approach. Food Chemistry. 2016;**210**:551-557. DOI: 10.1016/j.foodchem.2016.05.004

[40] Martínez Bueno MJ, Díaz-Galiano FJ, Rajski L, Cutillas V, Fernández-Alba AR. A non-targeted metabolomic approach to identify food markers to support discrimination between organic and conventional tomato crops. Journal of Chromatography. A. 2018;**1546**:66-76. DOI: 10.1016/j.chroma.2018.03.002

[41] Cichon MJ, Riedl KM, Schwartz SJ. A metabolomic evaluation of the phytochemical composition of tomato juices being used in human clinical trials. Food Chemistry. 2017;**228**:270-278. DOI: 10.1016/j.foodchem.2017.01.118

[42] Sidwick KL, Johnson AE, Adam CD, Pereira L, Thompson DF. Use of liquid chromatography quadrupole time-of-flight mass spectrometry and metabonomic profiling to differentiate between normally slaughtered and dead on arrival poultry meat. Analytical Chemistry. 2017;**89**(22):12131-12136. DOI: 10.1021/acs.analchem.7b02749

[43] Gallart-Ayala H, Kamleh MA, Hernández-Cassou S, Saurina J, Checa A. Ultra-high-performance liquid chromatography–high-resolution mass spectrometry based metabolomics as a strategy for beer characterization. Journal of the Institute of Brewing. 2016;**122**(3):430-436. DOI: 10.1002/jib.340

[44] Subbaraj AK, Kim YHB, Fraser K, Farouk MM. A hydrophilic interaction liquid chromatography-mass spectrometry (HILIC-MS) based metabolomics study on colour stability of ovine meat. Meat Science. 2016;**117**:163-172. DOI: 10.1016/j.meatsci.2016.02.028

Chapter 2

Large Molecule Fragmentation Dynamics Using Delayed Extraction Time-of-Flight Mass Spectroscopy

Najeeb Punnakayathil

Abstract

A time-of-flight mass spectrometer with a pulsed electron beam, delayed and pulsed extraction of the recoil ion is reported. This new technique is named as Delayed Extraction Time of Flight Mass Spectrometer (DEToF). The effectiveness of this technique is highlighted by studying the statistical decay of mono-cations over microsecond timescales from large molecules. Various details of the design and operation are performed in the context of electron impact ionization and fragmentation of few PAHs, naphthalene ($C_{10}H_8$), quinoline (C_9H_8N) and it isomer Isoquinoline (C_9H_8N) and are used as a test bench mark for large molecules fragmentation dynamics using DEToF. In this chapter we discuss the fragmentation dynamics of Naphthalene molecule and time evolution of various fragmentation channels of these PAH, explored using a rapid but delayed extraction of recoil ions. The temporal behavior of acetylene (C_2H_2), HCN, diacetylene (C_4H_2) and C_2H_2+HCN loss are observed and compared with the associated Arrhenius decay constant, internal energy and plasmon excitation energy.

Keywords: large molecule spectroscopy, delayed extraction time of flight, PAHs

1. Introduction

Large molecules in general and in particular, polycyclic aromatic hydrocarbons (PAHs) are found in the terrestrial environment as well as in the interstellar medium abundantly [1]. They are primarily formed on earth by the incomplete combustion of organic molecules. The origin of Diffuse Interstellar Bands (DIBs), that are absorption features seen in the spectra of astronomical objects in optical and infrared wavelengths has been attributed to PAHs [2]. Considering the abundance of high energy radiation in the interstellar medium, it remains an interesting endeavor to understand the mechanism behind the survivability of PAHs in such environments. Because of such astronomical significance, the interaction of PAHs with photons and charged particles has seen renewed interest in the last couple of decades. Even in recent times, several results have been reported globally on the topic of ion-PAH collisions [3–6]. In addition, detailed experiments are being carried out using synchrotron radiation sources [7]. High energy electron interaction with PAHs, on the other hand, has not been investigated in the context of radiation tolerance of PAHs.

High energy photon impact studies are often made using synchrotron radiation sources [8–10] with atoms or molecules as targets. Such mass spectrometric techniques along with the secondary electron selectivity methods like photoelectron-photoion coincidence (PEPICO) spectroscopy and threshold photoelectron-photoion coincidence (T-PEPICO) spectroscopy helps in determining the appearance energy and time scales of various dissociation channels of molecules very accurately. By modeling the line shape of the mass spectrum that arises due to slow decay, corresponding decay constant can be measured for microsecond range [11]. Longer decay times are probed using ion traps with variable extraction time [12]. It is essential to note that the excitation mechanism in conjunction with the appropriate electron spectrometer gives a very narrow range of internal energy left in the molecule. In particular, near the threshold, the internal energy is generally larger than the original thermal energy of the molecule, leading to resulting decay constants also range in a narrow band of values. Considering that the Arrhenius law decay rates are extremely sensitive to the internal energy, this factor very importantly implies that the decay rates with such secondary electron gated species will lie in a narrow range. Charged particle interaction with molecules will have a much broader range of internal energies and a precise value of decay constant cannot be obtained even with a suitable secondary electron energy gating. Hence modeling of exact decay curve for charged particle collision induced dissociation is deemed impractical. Moreover often minor fragmented peaks will interfere with the tail of mass spectrum due to isotopic effects or due to the presence of many hydrogen atoms in the molecule.

Electron impact ionization is one of the oldest mass spectrometric tools but it mainly focuses on identifying the possible ionization and fragmentation channels particularly between 70 and 100 eV energy. Typically, the mass spectrometric data available in the database is taken in this range, because ionization cross-section normally peaks in this region [13]. In past, several electron impact ionization investigations have been done mainly on inert gasses, diatomic or triatomic molecular gasses. Several experiments and modeling attempts have been made for such studies with electron energy up to few keV [14–16]. But such studies are very rare for larger molecules; the main reason is the complexity of a large number of decay channels, difficulties in separating indirect from direct ionization processes. For large molecules there are few attempts have been made in some specific cases for target specific energy loss modeling within the charged particle interaction with molecules [17].

The stability of PAHs during the interaction of charged particles, cosmic rays and photon sources in the interstellar medium is of our interest [4, 7]. It has been shown conclusively that for charged particle interaction with naphthalene, the plasmon excitation is a major channel particularly at high velocities of projectile wherein the other processes have diminishing cross section [4]. It is also seen that acetylene (C_2H_2) loss comes as a by-product of such plasmon excitation with a very specific range of decay constants. We use naphthalene as a model since it exhibits many general spectroscopic and structural properties of larger PAHs [18]. An attempt is made to investigate the interaction of high energy electron beam with PAHs and assess the time dependence of C_2H_2 evaporation in comparison with the other channels using a Time-of-Flight (ToF) mass spectrometer. Recent studies on benzene using PEPICO highlight the importance of the time variation of the decay channel [11]. We explore the time evolution of various decay channels within the first 5 μs of naphthalene ionization. This is achieved by the correlated pulsing of the electron source and recoil ion extraction field. We specifically target C_2H_2 and diacetylene (C_4H_2) evaporation process to identify the influence of plasmon excitation at different electron beam energies up to 1 keV.

The pulsed extraction of ions in a ToF setup can be used to analyze the evolution of a time-dependent population of various fragmentation channels. If the decay constants are in the range of $10^5 \sim 10^6$, then the thermal velocity dispersion does not affect the collection efficiency if it becomes possible to probe the system within sufficient time. Even though such a unique system has limited applicability, it has helped in our present investigation to study the evaporative loss from a PAH molecule due to electron impact and the decay rate for acetylene and diacetylene loss in naphthalene has been shown to have decay constant in the range of $10^5 \sim 10^6$ sec^{-1} at 8–9 eV internal energy [12, 18]. So, the decay time of acetylene and diacetylene is of the order of 1–10 μs. The ionization potential of naphthalene is 8.12 eV [19] and the plasmon resonance of naphthalene is observed at 17 eV [4]. The instrumentation presented here is very effective under such conditions.

2. Experimental set-up

For detection of product ions we use a Wiley-McLaren type time-of-flight mass spectrometer [20] with the pulsed extraction technique. The experimental set-up with the data acquisition system is shown in **Figure 1**. This experimental set-up consists of pusher (labeled as P_1 in **Figure 1**) as well as puller (labeled as P_2 in **Figure 1**) plates of thickness 1 mm and outer diameter of 105 mm and the puller plate has an opening diameter of 26 mm and for field uniformity that is covered with a nickel mesh characterized by 40 lines per inch, which allows a transmission efficiency of 95% of the molecular ions. For the field-free drift of ions, we have a drift tube of length 200 mm with an opening of 25 mm covered by a nickel mesh for field uniformity. Between pusher (P_1) and puller plates (P_2) there is 16 mm gap while that between the puller and the drift tube is 5 mm. A low current – high energy (1–5 keV) electron gun is used for ionizing the target molecule, which is custom made using CRT tube. Electron gun produces electron from filament via thermionic emission with a current of typically about 180 mA. There are two sets of

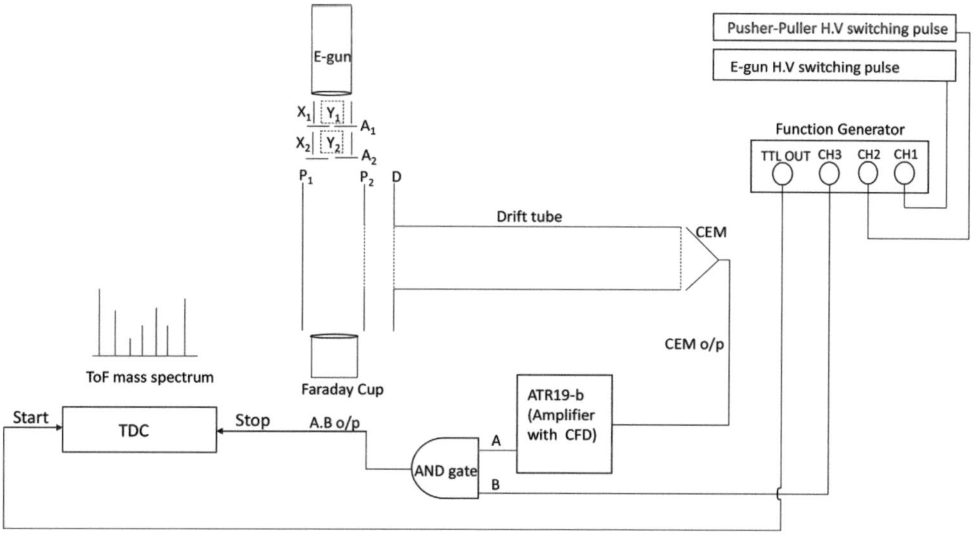

Figure 1.
The developed experimental set-up.

positional X-Y deflectors mounted after the focusing lens and two apertures (A_1, A_2), one between the two deflector sets (A_1) and the other after the deflectors (A_2), to discriminate the secondary electrons from the internal scattering of the beam. The whole electron gun assembly is floated, which voltage is decides the electron beam energy. A pulsed switch with a variable ON time width is used to operate the electron gun. For pulsed extraction ToF, a high voltage in the push-pull mode with a 50 ns rise and fall time is used which is a home-built high voltage MOSFET switch. A cylindrical Faraday cup made from stainless steel (SS-304), of length 100 mm and radius 25 mm to collect electron beam. This Faraday cup is placed at a distance of 50 mm from the exit of the interaction region. The Faraday cup is biased to +36 V using batteries to collect projectile electrons as well as the secondary electrons produced by the primary beam due to collision with Faraday cup wall. The intended target molecule is introduced into the interaction zone that is well localized in space through a fine capillary of internal diameter 300 microns and length of 15 mm. To avoid any possible secondary electron emission due to the electron beam colliding with the capillary, the capillary exit is kept nearly 5 mm away from the center of the ToF interaction region. A channel electron multiplier (CEM) is used for the detection of molecular ions, which is biased to a voltage of −2600 V. The electron gun, the pusher and puller plates are pulsed as per our pulsing sequence for delayed time-of-flight mass spectrometry as shown in **Figure 2**. Molecular ions produced due to electron impact (in the well-focused interaction region) are accelerated by the electric field and compensated for the special spread in the second region before entering the field-free drift tube followed by the ion detector. A multi-hit time-to-digital converter [Agilent TDC (Model: U1051A)] is employed as part of data acquisition system. To filter out the noises picked up by the detector due to switching pulses, we use a desired gate pulse.

Stainless steel (SS − 304) high vacuum chamber with metal joints are used to place the whole experimental setup. For better alignment, the time-of-flight mass spectrometer is mounted along the axis of the chamber and electron gun in perpendicular to its axis which ensure the cross beam (perpendicular) interaction of projectile beam and the molecular jet. There are several auxiliary ports for pumping, electrical connections and vacuum gauges. The whole chamber is pumped by two

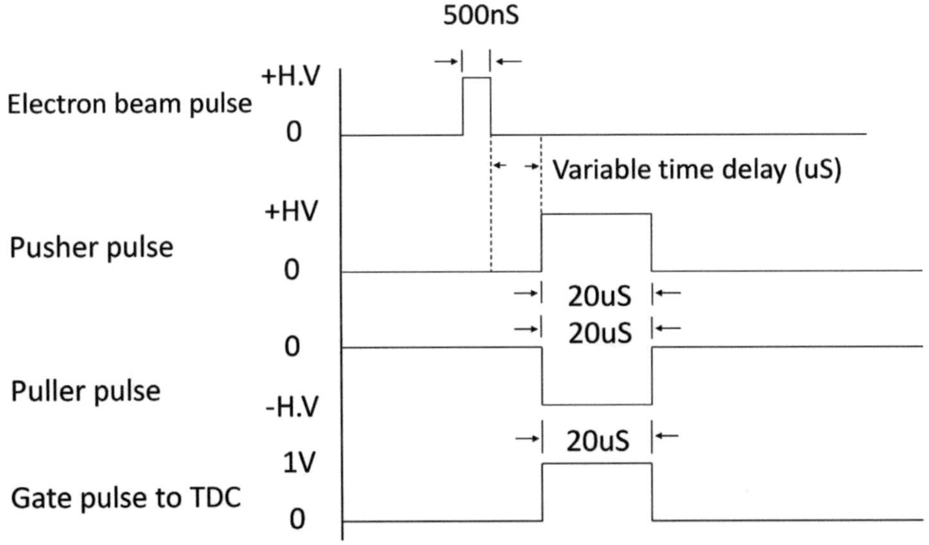

Figure 2.
Pulsing sequence used for delayed extraction time-of-flight mass spectrum.

turbo molecular vacuum pumps backed by a rotary pump. The pressure of the chamber goes down to 5.8×10^{-8} mbar, that rises up to a maximum of 5.0×10^{-7} mbar with the target gas. With naphthalene as the target, no heating was required due to the relatively large vapor pressure at 300 K. With the naphthalene target (has a purity greater than 98 %), the chamber pressure was maintained at $\sim 3.0 \times 10^{-7}$ mbar.

3. High voltage MOSFET switch

Electron beams are very sensitive to electrode voltages and hence in a time of flight spectrometer used in electron impact studies, they need special arrangement to limit the effect of the extraction voltages on the pusher and puller. Usually, this can be achieved by pulsing the extraction voltages. Switching noise pickup on the detector channels are the major complications with this arrangement. The electron beam cannot be allowed to persist during the extraction process, failing to which the beam might cause a large number of secondary electrons produced due to the deflected beam hitting the electrodes and then causing additional undesired ionization events. While we extracting the recoil ions it becomes essential to blank the electron beam. For this purpose, the present experiment with electron impact ionization requires two sets of fast switches, one set for switching electron beam and the other for switching of pusher and puller simultaneously. By using commercial off the shelf power MOSFETs we built fast switches to achieve this. Very commonly used fast single output switches are detailed in few literature for various applications [21–24]. Dual output fast pulsers for time-of-flight mass spectrometry are also available commercially, but are quite expensive generally.

It is necessary to have a pair of switches which can operate in a synchronized manner to switch the pusher and puller plates in the spectrometer which enables us to control the extraction cycle accurately. We use a pair of fast power MOSFETs triggered in synchronism to achieve this. A brief schematic of the high voltage MOSFET switch circuit is as shown in **Figure 3**. In order to minimize the probability of damage to the external TTL pulse generator due to noise produced by the high

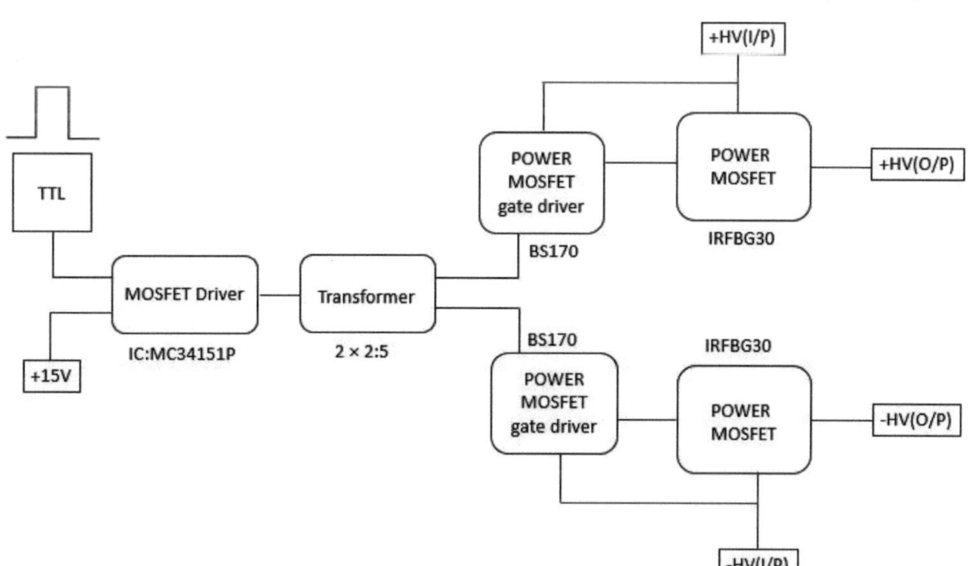

Figure 3.
Schematic of high voltage MOSFET switch (push-pull).

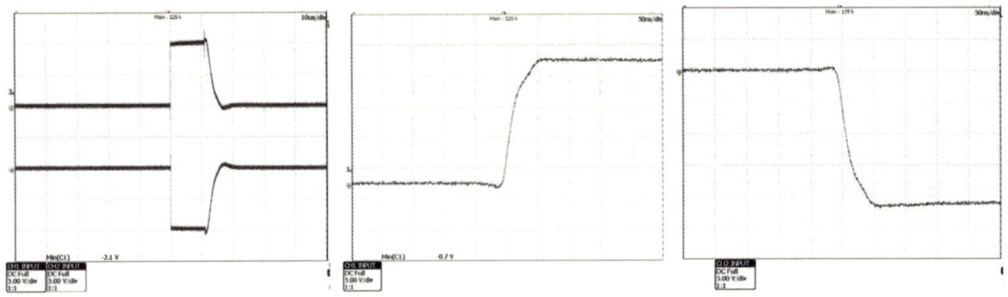

(a) (b) (c)

Figure 4.
High voltage MOSFET switch output (a) push-pull mode, (b) rise time ($<50nS$) in push mode, (c) fall time ($<50nS$) in pull mode.

voltage section of the circuit, the TTL pulse is passed through several logic gates. This pulse is then used to trigger a MOSFET driver (IC31415P) and the output of the driver is fed to a toroidal transformer with single primary (two turns) and a pair of secondary (five turns each). Gate of the main power MOSFET is controlled by each of the two outputs, triggers a pair of small signal MOSFETs (BS170) which in turn control the. We have tested the switch with voltages as high as ±300 V and could achieve rise/fall times of less than 50 ns with load (shown in **Figure 4b, c**). **Figure 4a** shows the fall/rise time at the end of the pulse was about 5 μs (note push-pull mode fall time) and was not necessary to improve as it played no role in the measurement.

A pair of deflectors and an aperture in between are used to achieve the electron beam pulsing. A single output MOSFET switch is used for one of the deflectors closer to the electron gun and this switch was derived from the same circuit described above but by using only one branch and using a cascade of four identical MOSFETs to achieve pulsing ability for voltages as high as ±2000 V.

4. Simulation

For optimized geometry as well as voltages for ion trajectory (as shown in **Figure 5a**, we performed the Simulation of our time-of-flight mass spectrometer with SIMION8.0 [25]. We simulated conditions with as much as 2.5 mm

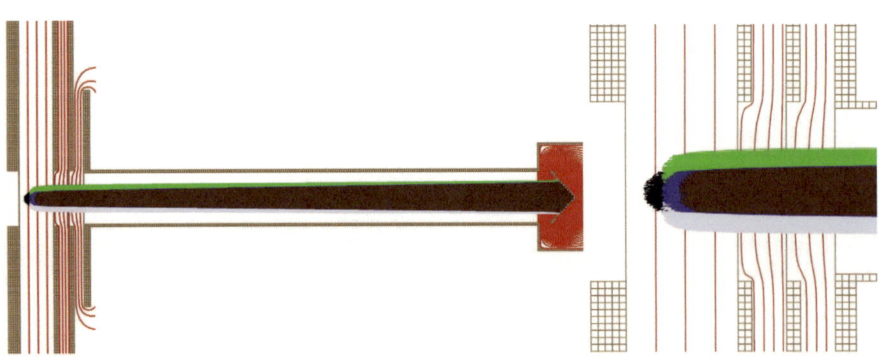

(a) (b)

Figure 5.
(a) Trajectory simulated for naphthalene ($C_{10}H_8$) with 480 μm/μs velocity from different position (colored as blue, green, black, white, brown) in the interaction region (b) Particle distribution in the interaction region. Red contours are the equipotential lines.

displacement of the center of the interaction region along the ToF axis as well as in the direction perpendicular to the ToF axis (**Figure 5b**). From our simulation the collection efficiency is estimated by assuming a spherical distribution of the source of diameter 6 mm and assuming an rms velocity twice as large as the value of 300 K that is 480 μm/μs is as the worst case scenario and in all the cases, we could achieve 100% collection efficiency. The electron beam is pulsed with a width of 500 ns. By using deflector blank pulsing method the pulse rise and fall times of 50 ns was obtained, where we pulse the deflector just before the first aperture and hence the electron pulse was effectively on for about 400 ns. The molecular ion extraction field (i.e., pusher-puller plates) is pulsed to 125 V/cm to extract the ions immediately after electron beam pulse is deflected off. The pusher-puller pulse are delayed for delayed extraction of molecular ions, for various delay times with reference to the electron beam pulse. From our calculation 99.9% of the naphthalene intact ions are expected to have thermal velocities less than 480 μm/μs at room temperature and this implies that they can move only 2.5 mm in 5 μs, as our simulations gave us the freedom of shifting the source position of the naphthalene target over the range of 2.5 mm. Thus, we have considered 5 microseconds as the maximum value of our delayed extraction time. The effect of the rise time of the extraction voltage to the molecular velocity spread is numerically calculated, which is 20 μm/μs for naphthalene molecule. So by taking into consideration the spread in thermal velocity, as well as the field effect for naphthalene molecule, the total spread in the velocity is expected to be a maximum of 500 μm/μs. In this case also we could achieve collection efficiency of 100% as per our simulation.

5. Result and discussions

Various projectile electron beam energy values with varying amounts of delay between the electron pulse-off and extraction pulse-on time are used for recording the Naphthalene mass spectra. The mass spectra obtained at different beam energies, as well as extraction delays, are systematically normalized to the single ionization peak area. For comparison between different beam energies and delay combinations, the area of each individual peak after such normalization could directly be considered.

A typical mass spectra is dominated by singly ionized naphthalene molecule followed by prominent peaks originating due to acetylene evaporation losses, as well as intact di-cation, and di-cation with $H_2/2H_2$ loss as shown in **Figure 6**. There are several energetic fragmentation channels are also visible but are not as prominent. The intensity of single C loss and minimal intensity of 3C loss are hardly seen. The single ionization peak is preceded by one or two H atom loss is in general. The mass spectra reported here is not discernible due to our modest resolution. The possible loss of multiple H along with acetylene loss is negligible as shown in the acetylene loss region. On the other hand, the diacetylene loss region, shows two clear peaks due to loss of C_4H_2 and C_4H_4 and the latter could be due to the sequential loss of two acetylene molecules. Significantly, the peak following mass at 64 mass units can be assigned clearly to the di-cation. The formation of exactly half mass fragment for naphthalene is highly improbable unless it is accompanied by several other peaks due to the concurrent loss of multiple H atoms and that would cause a large spread in the mass spectra. A dominant $2H/H_2$ loss channel is observed with about 50% of the di-cation peak intensity and a fast evaporation of neutral 2H or H_2 and hence a large internal energy deposition due to low impact parameter collision is indicated distinct appearance of such a peak. We could ignore safely the 2H loss channel as it is observed to be less likely in previous investigations as demonstrated by Jochims et al. [8] and thus unless stated explicitly otherwise, we will only refer to H_2 loss possibility wherever the loss of 2 mass units is concerned. The effect is even stronger in the case of di-cation that a large peak is observed at 51 mass units coming from di-cation of naphthalene with the loss of one acetylenes and the acetylene loss competes with H loss in the case of single ionization. Similarly, this peak is again accompanied by H_2 loss along with acetylene loss. As shown earlier in ion-PAH collisions studies [4, 6] we see the dominant H_2 channel. It is $C_3H_3^+$ and $C_3H_4^+$ that dominate the spectra and the rest largely come as by-products of various decay channels in the mass region near 39 units.

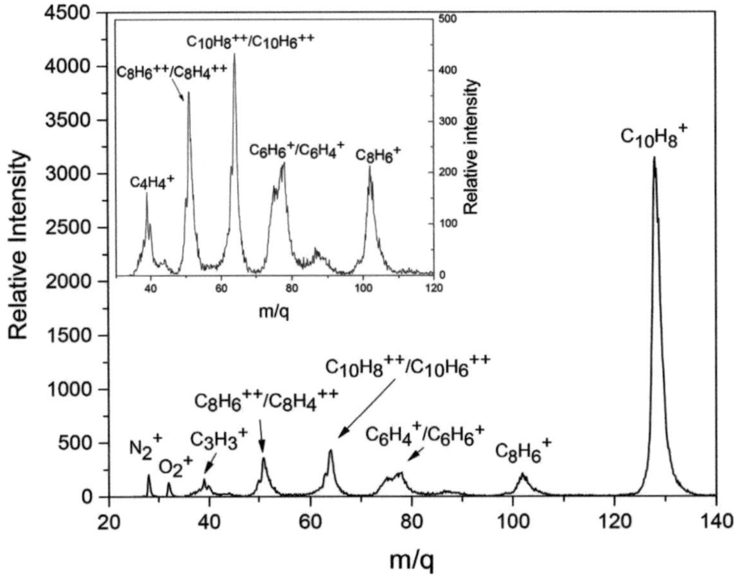

Figure 6.
The mass spectrum of naphthalene($C_{10}H_8$) at 1000 eV electron impact (Inset: naphthalene fragments).

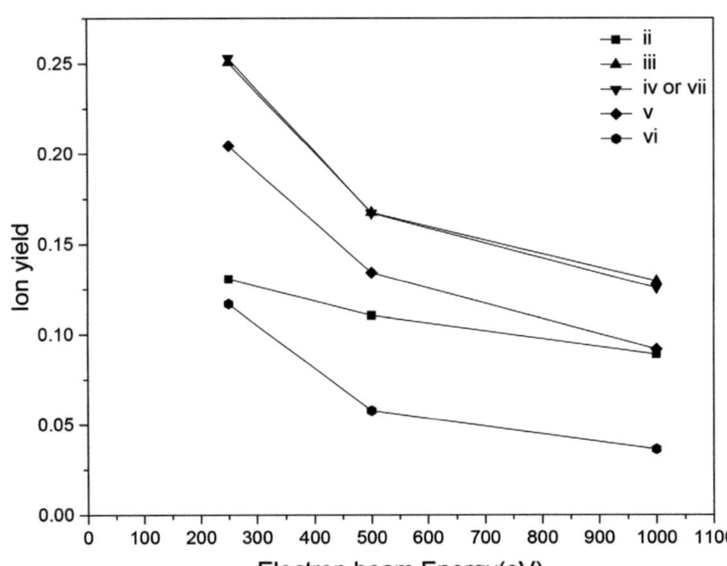

Figure 7.
Projectile electron beam energy dependence of different decay channel with zero delay time extraction. Decay channels are labeled as listed above.

For low-Z targets like hydrocarbons, the projectile energy dependence of the electron impact ionization process is very commonly known to peak at about 70–100 eV and at higher energy, the cross section varies as $\ln E/E$. Interestingly, the acetylene loss channel that shows a distinctly different behavior, as seen from **Figure 7** except that all the channels considered here follow a similar trend.

Thus, we have considered the following decay channels for our analysis.

i. $$C_{10}H_8 + e^- \rightarrow C_{10}H_8^+ + 2e^-$$

ii. $$C_{10}H_8 + e^- \rightarrow C_8H_6^+ + C_2H_2 + 2e^-$$

iii. $$C_{10}H_8 + e^- \rightarrow C_8H_5^+ + C_2H_2 + H + 2e^-$$

iv. $$C_{10}H_8 + e^- \rightarrow C_6H_6^+ + C_4H_2 + 2e^-$$

v. $$C_{10}H_8 + e^- \rightarrow C_6H_5^+ + C_2H_2 + C_2H + 2e^-$$

vi. $$C_{10}H_8 + e^- \rightarrow C_6H_4^+ + C_4H_4(2 \times C_2H_2) + 2e^-$$

vii. $$C_{10}H_8 + e^- \rightarrow C_5H_3^+ + C_2H_2 + C_3H_3 + 2e^-$$

viii. $$C_{10}H_8 + e^- \rightarrow C_4H_4^+ + C_4H_2 + C_2H_2 + 2e^-$$

ix. $$C_{10}H_8 + e^- \rightarrow C_4H_3^+ + 2 \times C_2H_2 + C_2H + 2e^-$$

x. $$C_{10}H_8 + e^- \rightarrow C_4H_2^+ + 3 \times C_2H_2 + 2e^-$$

xi. $$C_{10}H_8 + e^- \rightarrow C_3H_3^+ + C_3H_3 + C_4H_2 + 2e^-$$

xii. $$C_{10}H_8 + e^- \rightarrow C_{10}H_8^{++} + 3e^-$$

xiii. $$C_{10}H_8 + e^- \rightarrow C_{10}H_6^{++} + 2H/H_2 + 3e^-$$

xiv. $$C_{10}H_8 + e^- \rightarrow C_8H_6^{++} + C_2H_2 + 3e^-$$

xv. $$C_{10}H_8 + e^- \rightarrow C_8H_4^{++} + C_2H_2 + 2H/H_2 + 3e^-$$

The loss of one or more H atom from the intact mono-cation is not shown explicitly in the list but is one of the most important channels. As observed by Gotkis et al., the primary decay channel for naphthalene is either H loss or acetylene loss [12]. It should be noted that the loss of multiple H and acetylene fragments are the prominent statistical decay modes for PAHs and therefore they can be very useful in understanding the dynamics of the internal degrees of freedom in PAHs. Acetylene loss is clearly the most useful channel to study from the data we present here. Higher order ionizations like double and triple ionization are low impact parameter processes and thus a detailed direct coulomb interaction model is more relevant in such cases. In this study our main goal is to explore the plasmon excitation which is a large impact parameter process and it is known to produce singly charged cations for the case of PAHs [26]. We expect strong dynamical and statistical effects in the singly charged naphthalene ions and the associated evaporation products like single and double acetylene loss in our case. Various decay channels are referred in the manuscript according to the numbers given in the list above.

Jochims et al. are discussed the reaction sequence for acetylene loss and the formation of phenyl-acetylene $C_8H_6^+$ from $C_{10}H_8$ in detail [26]. An intermediate is formed by the breaking of the transannular bond accompanied by the migration of H in the naphthalene cation and this is followed by the successive cleavage of C-C bond with a loss of C_2H_2 molecule and the formation of polyacetylene cation [26].

The first six decay channels are considered here with mono-cations and the rest may come from very energetic mono-cation or di-cation. Thus the processes governing the production of the latter decay channels can be treated as low impact parameter and high internal energy channels. And subsequently these decay channels are expected to have substantially large decay rates compared to the decay rates for mono-cations leading to decay channels ii and iii. This unique behavior is evident as shown in **Figure 8a–c** in which plot the relative intensity of all the decay channels at 250, 500 and 1000 eV electron impact energies. The very first clear observation is that the decay channels ii evolve very differently compared to the rest of the channels as a function of extraction delay and this can be seen in **Figure 9a, b**. The diacetylene loss channel show an increase in the yield fraction followed by a steady population or a slight decrease up to 5 μs extraction delay on the other hand, the last five decay channels show a clear decrease from the zero delay onwards. This decay behaviors indicates that the last five reactions have decay rates that are much larger than 10^6 and assuming these arise from the statistical dissociation.

The interaction energy for the formation of larger sized fragments according to reactions iv - viii needs to be well above 22 eV as discussed by Ruhl et al. [18]. The plasmon excitation is expected to peak at about 17 eV [26]. In the data for each beam energy, we observe that compared to the acetylene loss the decay channel iii peak shows a faster rate of change and this can be understood from the fact that the decay rate of channel iii is marginally higher than that of the acetylene loss channel. Interestingly we see a gradual decrease in the relative intensity of all the daughter channels as a function of projectile energy. We can assign two major possibly for the decay channel iii: [1] neutral acetylene evaporation from mono-cation and [2] the dissociation of $C_4H_2^+$ fragment. Moreover, this could also be the C_4H_n fragment but it is not resolved well in our spectra. The population of evaporative products grows substantially in earlier times and after about 3 μs, it starts to drop again and behavior is seen even at 500 and 1000 eV beam energies. This suggesting that the first increase comes due to plasmon excitation and reduces its contribution as we go to higher beam energies and this observation is in complete agreement with the behavior seen in decay channel ii. On the other hand, the rest of the channels, show gradual and continuous decrease in yield with extraction delay time. In the case of decay channel ii, the scales are longer than that for channel ii is seen. The decay channel iv to viii represents more energetic processes and hence show a decrease in

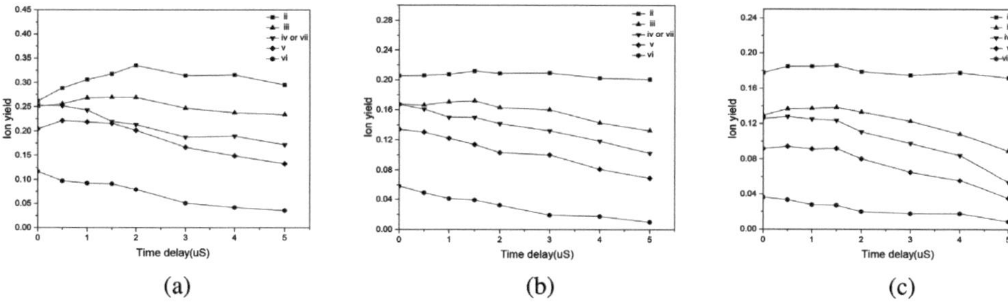

Figure 8.
Various fragmentation channels of naphthalene at (a) 250 eV, (b) 500 eV, (c) 1000 eV electron impact energies. Decay channels are labeled as listed above.

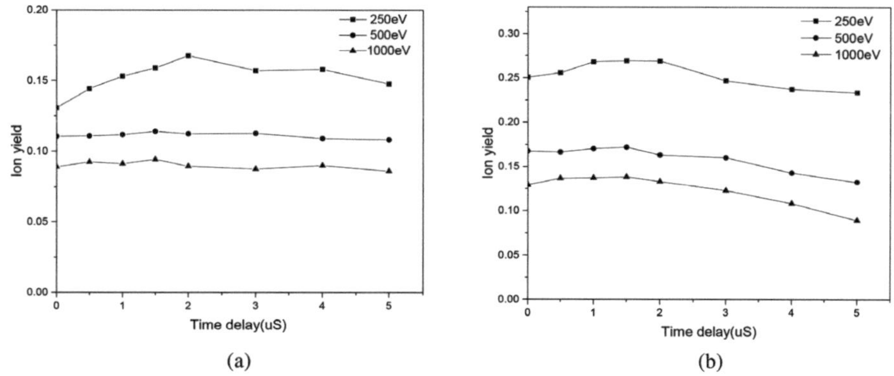

Figure 9.
(a) Decay channel ii (C_2H_2 Loss) and (b) decay channel iii (C_4H_2 or $2 \times C_2H_2$ loss) at various electron impact energies.

population fraction. These channels from the zero delay time, indicating sufficiently fast decay accompanied by sufficient kinetic energy release to cause fast dispersion of daughter ions from the interaction volume.

6. Conclusion

For PAHs in general and for naphthalene in particular a strong influence of the plasmon resonance excitation is known to be present. This effect has recently gained importance due to its possible role in the formation of molecular hydrogen and acetylene molecules in ISM and has been under investigation using far-UV photo-excitation as well as heavy ion-induced excitation. In this work, we have shown the effect of plasmon excitation under by high energy electron impact excitation. The time evolution of the acetylene evaporation, which is known to be a by-product of the plasmon excitation process, is measured for this study. A pulsed electron source along with the pulsed extraction of recoil ions using fast high voltage pulses is implemented and varying the extraction delay we observe the parent and daughter ion yields of naphthalene molecule is observed.

The most dominant low energy channels commensurating with the excitation energy, in range of 7–8 eV leading to a total energy loss of 15–16 eV range is the decay channels leading to loss of acetylene and the loss of diacetylene. Acetylene loss shows much weaker energy dependence compared to the other channels which again is a well-known property of plasmon excitation. In the case of acetylene evaporation processes the time scales of few microseconds are seen and yields of acetylene evaporation population grow by about 20% as the delay is varied from 0 to 5 μs. Importantly, this work demonstrates that the relatively simple and well-known technique of delayed extraction time-of-flight mass spectrometry can be very useful in certain circumstances like broader energy loss mechanisms i. e, plasmon excitation in charged particle interaction with PAHs. The time evolution of a group of statistical decay channels that could otherwise be inaccessible due to the large range of decay constants can be probed using this unique technique.

Author details

Najeeb Punnakayathil
Department of Physics, Stockholm University, Stockholm, Sweden

*Address all correspondence to: najeeb.punnakayathil@fysik.su.se

References

[1] Salama F, Allamandola LJ. Electronic absorption spectroscopy of matrix-isolated polycyclic aromatic hydrocarbon cations. i. the naphthalene cation ($C_{10}H^{+}{}_{8}$). The Journal of Chemical Physics. 1991;**94**(11): 6964-6977

[2] Lessen D, Freivogel P, Maier JP. Laboratory evidence for highly unsaturated hydrocarbons as carriers of some of the diffuse interstellar bands. Nature. 1993;**366**(6454):439-441

[3] Postma J, Bari S, Hoekstra R, Tielens AGGM, Schlathölter T. Ionization and fragmentation of anthracene upon interaction with kev protons and particles. The Astrophysical Journal. 2010;**708**(1):435

[4] Mishra PM, Rajput J, Safvan CP, Vig S, Kadhane U. Electron emission and electron transfer processes in proton-naphthalene collisions at intermediate velocities. Physical Review A. 2013;**88**: 052707

[5] Kadhane U, Misra D, Singh YP, Tribedi LC. Effect of collective response on electron capture and excitation in collisions of highly charged ions with fullerenes. Physical Review Letters. 2003;**90**:093401

[6] Reitsma G, Zettergren H, Boschman L, Bodewits E, Hoekstra R, et al. Ion-polycyclic aromatic hydrocarbon collisions: Kinetic energy releases for specific fragmentation channels. Journal of Physics B: Atomic, Molecular and Optical Physics. 2013;**46**(24):245201

[7] Mishra PM, Avaldi L, Bolognesi P, Prince KC, Richter R, Kadhane UR. Valence shell photoelectron spectroscopy of pyrene and fluorene: Photon energy dependence in the far-ultraviolet region. The Journal of Physical Chemistry A. 2014;**118**(17): 3128-3135

[8] Jochims H, Baumgärtel H, Leach S. Structure-dependent photostability of polycyclic aromatic hydrocarbon cations: Laboratory studies and astrophysical implications. The Astrophysical Journal. 1999;**512**(1):500

[9] Boschi R, Clar E, Schmidt W. Photoelectron spectra of polynuclear aromatics. iii. The effect of nonplanarity in sterically overcrowded aromatic hydrocarbons. The Journal of Chemical Physics. 1974;**60**(11):4406-4418

[10] Schmidt W. Photoelectron spectra of polynuclear aromatics. v. correlations with ultraviolet absorption spectra in the catacondensed series. The Journal of Chemical Physics. 1977;**66**(2): 828-845

[11] Alagia M, Candori P, Falcinelli S, Pirani F, Mundim MSP, Richter R, et al. Dissociative double photoionization of benzene molecules in the 26–33 ev energy range. Physical Chemistry Chemical Physics. 2011;**13**(18): 8245-8250

[12] Gotkis Y, Oleinikova M, Naor M, Lifshitz C. Time-dependent mass spectra and breakdown graphs. 17. naphthalene and phenanthrene. The Journal of Physical Chemistry. 1993; **97**(47):12282-12290

[13] Linstrom PJ, Mallard WG. NIST Chemistry WebBook, NIST Standard Reference Database Number 69. Vol. 20899. Gaithersburg, MD: National Institute of Standards and Technology; 2016

[14] Deutsch H, Margreiter D, Märk T. A semi-empirical approach to the calculation of absolute inner-shell electron impact ionization cross sections. Zeitschrift für Physik D Atoms, Molecules and Clusters. 1994; **29**(1):31-37

[15] Kim Y-K, Rudd ME. Binary-encounter-dipole model for electron-impact ionization. Physical Review A. 1994;**50**:3954-3967

[16] Vallance C, Harland PW, Maclagan RG. Quantum mechanical calculation of maximum electron impact single ionization cross sections for the inert gases and small molecules. The Journal of Physical Chemistry. 1996;**100**(37): 15021-15026

[17] Hwang W, Kim Y-K, Rudd ME. New model for electron-impact ionization cross sections of molecules. The Journal of Chemical Physics. 1996; **104**(8):2956-2966

[18] Ruehl E, Price SD, Leach S. Single and double photoionization processes in naphthalene between 8 and 35 ev. The Journal of Physical Chemistry. 1989; **93**(17):6312-6321

[19] Van Brunt RJ, Wacks ME. Electron-impact studies of aromatic hydrocarbons. iii. Azulene and naphthalene. The Journal of Chemical Physics. 1964;**41**(10):3195-3199

[20] Wiley WC, McLaren IH. Time-of-flight mass spectrometer with improved resolution. Review of Scientific Instruments. 1955;**26**(1150):39-441

[21] Baker R, Johnson B. Series operation of power mosfets for high-speed, high-voltage switching applications. Review of Scientific Instruments. 1993;**64**(6): 1655-1656

[22] Jiang W. Fast high voltage switching using stacked mosfets. IEEE Transactions on Dielectrics and Electrical Insulation. 2007;**14**(4): 947-950

[23] Chappell P, Campden K. Switching performance of power mosfets with capacitive loads at high frequency and high voltage for square wave generators. Measurement Science and Technology. 1992;**3**(4):356

[24] Bernius MT, Chutjian A. High-voltage, full-floating 10-mhz square-wave generator with phase control. Review of Scientific Instruments. 1989; **60**(4):779-782

[25] Dahl D. Simion 3D Version 6.0. In: 43rd ASMS Conference on Mass Spectrometry and Allied Topics; Atlanta, Ga; 1995. p. 717

[26] Jochims H, Rasekh H, Rühl E, Baumgärtel H, Leach S. The photofragmentation of naphthalene and azulene monocations in the energy range 7–22 ev. Chemical Physics. 1992; **168**(1):159-184

Chapter 3

Mass Spectrometry as a Workhorse for Preclinical Drug Discovery: Special Emphasis on Drug Metabolism and Pharmacokinetics

Vijayabhaskar Veeravalli, Lakshmi Mohan Vamsi Madgula and Pratima Srivastava

Abstract

Mass spectrometry as an instrument is popular, given its sensitivity, selectivity, speed and robustness. In this chapter, we have briefly deliberated on various mass spec platforms, their hardware components and specific applications in preclinical drug discovery with a special emphasis on drug metabolism and pharmacokinetic assays. Basic principle of operation of mass spectrometer and various ionization techniques/mass analyzers was explicitly discussed. Compatibility of mass spectrometers with ultrafast LC and various throughput techniques, enabled evaluation of thousands of compounds with quick turnaround times. Faster generation of results corresponding to in vitro ADME and in vivo pharmacokinetic assays, aid medicinal chemists to refine their combinatorial synthetic chemistry efforts and expedite the lead optimization and identification phases of drug discovery. Mass spectrometer is a powerful tool for both qualitative and quantitative applications. While quantitative applications include measurement of absolute/relative concentrations, qualitative features assist in identification of molecular structures of metabolites and putative biotransformation pathways. Qualitative inputs are more precise and accurate, with the advent of high-resolution mass spectrometry technology. Although, mass spectrometry has many built-in advantages, it also suffers from matrix effects, as the samples analyzed are mostly of biological origin and are complex in nature. In this chapter, we have defined the nature of matrix effects and various approaches by which these matrix effects can be mitigated.

Keywords: mass spectrometer, mass analyzers, ionization sources, DMPK, ADME, LC–MS/MS, matrix effects

1. Introduction

Since its discovery 100 years ago by Sir J.J. Thompson for the quantitative measurement of the mass and charge of cathode rays, mass spectrometer eventually evolved as a reliable analytical platform aimed at the analysis of small and large molecules [1]. While hyphenation of mass spectrometry with gas chromatography achieved its early success, however, liquid chromatography could not due to indigent mass spec interfaces. This is given the inability of interfaces to handle higher flow rates of liquid sample. Additionally, ionization techniques such as chemical and electron impact ionization did not suit for thermolabile and high molecular weight compounds [2, 3]. Advancement in ionization techniques such as fast atom bombardment that suited for analysis of large molecules, thermospray and particle beam ionization which were efficient for small molecules enabled the fruitful hyphenation of liquid chromatography with mass spectrometry [4–7]. Thermospray in general forms ammonium adducts, while particle beam generated electron impact spectra. Within a few years thermospray was succeeded by atmospheric pressure ionization techniques such electrospray ionization (ESI), atmospheric pressure chemical ionization (APCI), atmospheric pressure photoionization (APPI) and atmospheric pressure matrix assisted laser desorption ionization (AP-MALDI). Mass spectrometer operates on the principle of ionization of analytes followed by their separation based on their mass-to-charge ratio [8]. Various mass spectrometer analyzers ranging from linear trap, ion trap, triple quadrupole, time of flight and orbitrap have specific applications for sample analysis in preclinical drug discovery and development.

Medicinal chemistry efforts in drug discovery is majorly focused on understanding the right combination of new chemical entity (NCE) properties that helps in cherry-picking the compounds with promising properties to progress from discovery to development phase. In the process of lead optimization, NCE's are subjected to a series of drug metabolism and pharmacokinetic assays to assess the druggable properties and mitigate late stage failures. Almost 40% of failures in development phase were due to poor pharmacokinetic properties of NCE's [9]. However, this percentage had gradually decreased to 10% as major pharmaceutical companies incorporated drug metabolism and pharmacokinetic screening (DMPK) in lead optimization phase of drug discovery. In this view, compounds must pass through a series of screens that scrutinize the problematic compounds until a small number have been selected for more rigorous testing in the development phase. Hence, lead optimization typically is an iterative process that uses the DMPK data to optimize the druggable properties of NCE's. Regardless of the screening panel, qualitative and quantitative analytical results to understand absorption, distribution, metabolism and excretion (ADME) properties were generated using liquid chromatography–tandem mass spectrometry (LC–MS/MS) [10–17]. In this chapter, we have briefly discussed on various atmospheric pressure ionization techniques/mass analyzers, and applications of mass spectrometer in drug discovery with special emphasis on drug metabolism and pharmacokinetic assays.

2. Atmospheric pressure ionization

Ionization in atmospheric ionization sources occur at atmospheric pressure and ions then gets transferred into the vacuum. As the liquid completely converts in to gas phase in the ionization source, those ionization techniques that use atmospheric pressure ionization are more convenient to hyphenate with liquid chromatography. These ionization techniques include electrospray ionization (ESI), atmospheric pressure chemical ionization (APCI), and atmospheric pressure photoionization (APPI) and are most widely used. Additionally, MALDI that also uses atmospheric pressure ionization is getting popular with its unique feature in performing mass spectrometry imaging (MSI) and analysis of large molecules [18]. Also, other ionization techniques such as desorption electrospray (DESI) or direct analysis in real time (DART) are becoming popular for the analysis of surface or solid samples [19, 20]. However, their applications in the field of drug metabolism and pharmacokinetics are very limited.

2.1 Electrospray ionization (ESI)

In ESI, analytes initially get charged with the assistance of electrical energy and charged ions transfer from solution to gaseous phase, before subjecting to mass spectrometric analysis. Ionic species in solution can be analyzed as such, whereas neutral compounds can be converted to ionic species and studied by ESI-MS. Electrospray ionization occurs in four stages: 1) charging of analytes in the capillary tube 2) formation of fine spray of charged droplets 3) solvent evaporation 4) columbic explosion/Rayleigh scattering of ions from the droplet (**Figure 1**). The liquid effluent moves from liquid chromatography to the mass spectrometer through a fused silica capillary maintained at voltage of 2.5–6.0 KV. In negative mode, to avoid discharge the range is lower (3–4 KV) than positive mode. ESI is a condensed phase ionization process and the ions have to be already present in solution. To generate ions, the pH has to be adjusted in such a way that ionizable groups are either protonated or deprotonated. In some cases, neutral molecules can be analyzed by the formation of adducts with ions such as ammonium, sodium, potassium, acetate or silver. Charged droplets undergo nebulization in the presence of nebulizer gas. After nebulization, charged droplets further reduce in size with the assistance of heat and breakdown in to minute droplets. Finally, as the droplets grow smaller and smaller, ions get released in to gaseous phase by a mechanism called rayleigh scattering/columbic explosion. The emitted ions are sampled by a sampling skimmer cone and are then accelerated into the mass analyzer for subsequent measurement of molecular mass and ion intensity [21–24]. An important characteristic of ESI-MS is it works as a concentration-dependent detector, which means MS response is directly proportional to concentration of analyte. Hence, irrespective of flow rate of mobile phase post column to the ionization source, response remains the same as long as the source-gas conditions are optimal for the flow rate. ESI technique is suitable for the analysis of polar to moderately polar molecules.

2.2 Atmospheric pressure chemical ionization (APCI)

In atmospheric pressure chemical ionization, unlike ESI, sample evaporation occurs first, followed by ionization in gas phase through corona discharge needle (**Figure 2**) [25]. The ionization principle is mostly similar to chemical ionization; however, it occurs at atmospheric pressure. APCI-MS can also be called as mass-sensitive detector, as the higher flow that goes in to ionization source, the higher will

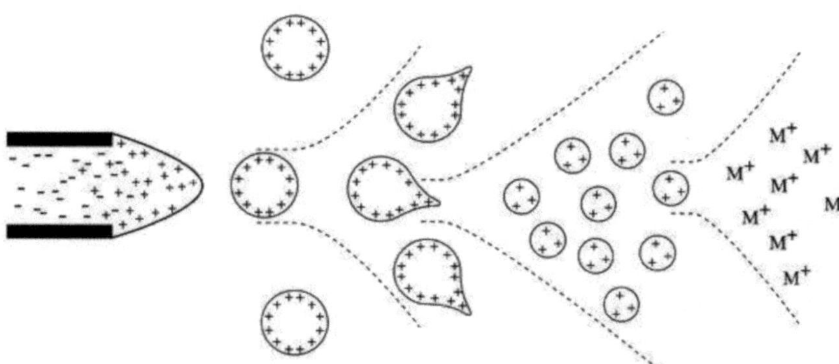

Figure 1.
Mechanism of electrospray ionization process.

be the peak response. One important requirement in APCI for optimal sensitivity is the sample has to be completely evaporated, before subjecting to ionization. First an aerosol of mobile phase is formed with the assistance of nebulizer gas. This aerosol is further subjected to heating at 200–550°C in a ceramic tube enabling complete evaporation. Even though higher temperatures are employed, the actual temperature felt by analyte molecules is way lesser due to a phenomenon called evaporative cooling effect/evaporation enthalpy. Next, analyte molecules in gas phases were bombarded with electrons formed from corona discharge needle [26, 27]. In positive mode primary ions such as $N2^+$ are formed by electron impact. These ions further react with water in several steps by charge transfer to form $H3O^+$. Ionization of the analytes occurs then by proton transfer from $H3O^+$. In negative mode ions are formed either by: (i) resonance capture (AB to AB^-), (ii) dissociative capture (AB to B^-) or (iii) ion–molecule reaction (BH to B^-). One disadvantage with APCI when compared to ESI is, APCI is not suitable for thermolabile compounds as typical temperatures experienced by the analyte molecules are ~150°C. However, in case of ESI, molecules encounter temperatures ~40°C in the process of evaporation.

2.3 Atmospheric pressure photo ionization (APPI)

APPI by design is similar to APCI, with only difference being the replacement of corona discharge needle with gas discharge krypton lamp (10.0 eV) that produces ultraviolet photons [28–30]. Evaporation of liquid phase happens in pneumatic nebulizer. While ionization potential for most of the analytes is less than 10 eV; mobile phase constituents such as water, acetonitrile and methanol has higher potential requirements. Presence of dopants such as toluene or acetone will help in enhancing the sensitivity of analyte ions. Dopant molecules absorb photon energy and eject an electron, resulting in the formation of radical cation. Ionization of analytes can happen by two processes: 1) Charge transfer between analyte and radical cations generated from dopant molecules. 2) Charge transfer between dopant molecules and mobile phase components and finally from mobile phase components to analytes (**Figure 3**). Similar to other atmospheric pressure ionization techniques, APPI is also suitable for negative mode of ionization. Sensitivity of APPI is flow rate dependent and better sensitivities have been reported at low flow rates. When compared to APCI, APPI offers lesser matrix effects and minimal source contamination. Success of APPI as an ionization technique was reported in the analysis of steroids and quinones [31–33].

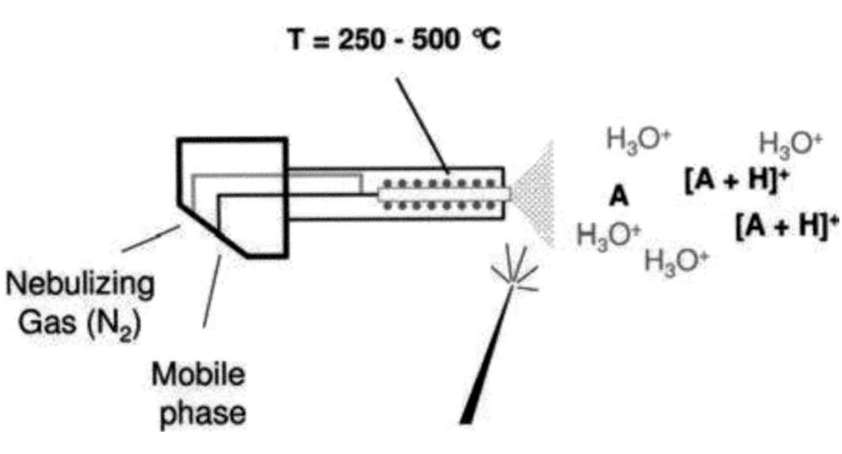

Figure 2.
Mechanism of atmospheric pressure chemical ionization process.

In a nut shell, ESI is useful for the analysis of moderately polar to highly polar analytes; APCI covers moderately polar to non-polar analytes, whereas APPI suits for non-polar to moderately polar analytes.

2.4 Matrix assisted laser desorption ionization (MALDI)

MALDI operates on principle of ionization of analytes dissolved in a matrix consisting of organic compound (sinapinic acid, α-cyano-4-hydroxycinnamic acid, 2,5-dihydroxybenzoic acid) and evaporated to dryness on a target plate [34]. Matrix crystallizes up on drying and the analyte dissolved with in it also gets co-crystallized. Firstly, laser beam hits the dried sample and ionize organic compound, which later ionize the analyte molecules. Laser beam causes both desorption and ionization of analytes. Nitrogen laser emitting at 337 nm and Nd: YAG laser emitting at 355 nm are the most widely used ones (**Figure 4**). MALDI is considered as a very

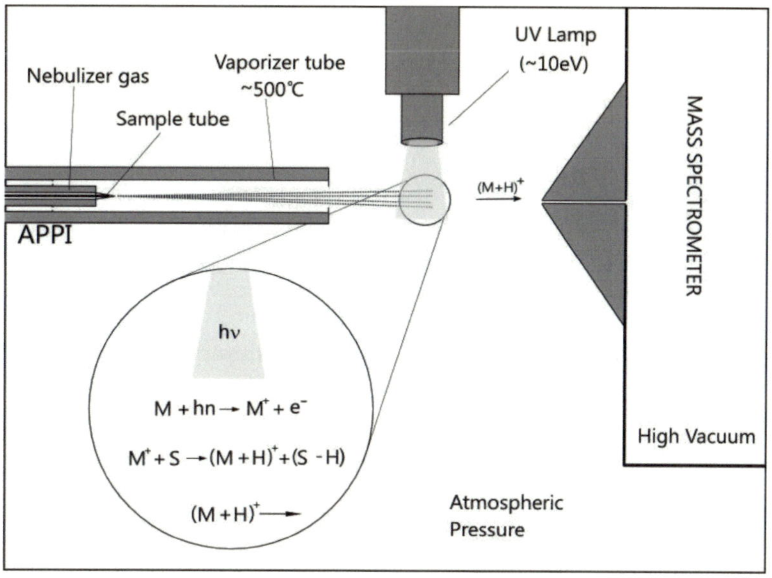

Figure 3.
Schematic representation of atmospheric pressure photo ionization process.

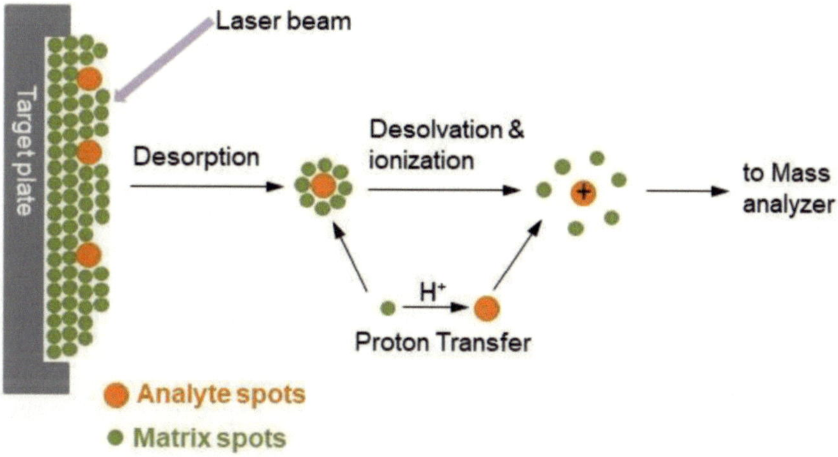

Figure 4.
Mechanism of matrix assisted laser desorption ionization process.

soft ionization technique that causes minimal fragmentation of the analyte ions and is also suitable for analysis of large molecules ranging from peptides to proteins, lipids and polymers [35–38]. It is also amenable to high throughput and target sample plates can be readily stored for future use. One major advantage of MALDI-MS is chromatographic separation of analytes is not required. However, due to lack of separation, matrix interferences impact the analytical results [39, 40]. Additionally, MALDI is not suitable for low molecular weight compounds. Also, MALDI needs TOF as an analyzer to cover high mass range in a linear mode, whereas ESI can be coupled with any mass analyzer. Recently, MALDI has been coupled to triple quadrupole and successfully used for the analysis of small molecules [41, 42].

3. Mass analyzers

3.1 Quadrupole analyzer

Quadrupole mass analyzer consists of four hyperbolic or circular rods positioned in parallel and are located diagonally at identical distances from each other. The rods are diagonally connected. Positive direct current (DC; U) is applied to one pair of rods and negative potential is applied to the other pair of rods. Apart from direct current, alternating radiofrequency (RF; $V\cos \omega t$) potential is also applied to these rods. The ion trajectory is affected in x and y directions by the total electric field composed by a quadrupolar alternating field and a constant field. Because there is only a two-dimensional quadrupole field the ions accelerated after ionization, maintain their velocity along the z axis.

The motion of ions in the quadrupole can be best described by Mathieu Equations [43]. The ions supposedly travel in a stable trajectory and only those ions that travel in stable trajectory reaches detector. Mass spectrum is obtained by ramping RF and DC voltages in a constant ratio. When DC voltage is set to zero and RF voltage is maintained, all ions pass through quadrupole. It is the DC voltage that helps in filtering out the ions of interest and generate mass spectrum (**Figure 5**). Hence, the quadrupoles that apply only RF voltages just act as ion guides or collision cell. Mass resolution for typical quadrupole analyzers falls in the range 0.6–0.8 da units, which is defined to be a unit resolution. However, current generation high resolution mass spectrometers offer to determine masses within 5–10 ppm error.

3.2 Triple quadrupole mass analyzer

Triple quadrupole mass analyzer consists of two RF/DC mass analyzers and two RF only mass analyzers. Q0and Q2 (collision cell) were considered to be RF only quadrupoles, whereas Q1 and Q3 falls under RF/DC mass analyzers [44]. Hence, Q0 and Q2 acts as ion guides and Q1 and Q3 acts as mass filters. Q0 acts as an ion guide by focusing all the ions obtained from ionization source to Q1. Q1 even though a RF/DC mass analyzer, can also be operated in RF only quadrupole depending on the type of analysis. When it comes to qualitative analysis, Q1 acts as a RF only quadrupole, whereas in case of quantitative analysis it acts as RF/DC quadrupole. Similarly, Q3 also operates in both modes based on the analytical requirements. Q2 in addition being a RF only quadrupole, acts as a collision cell to fragment the ions and generate compound specific information, which enables the mass spec to be a more specific and selective detection system (**Figure 6**). Process of generation of fragment ions in the collision cell is termed as collision induced dissociation (CID), which happens with the assistance of neutral argon or nitrogen gas [45, 46].

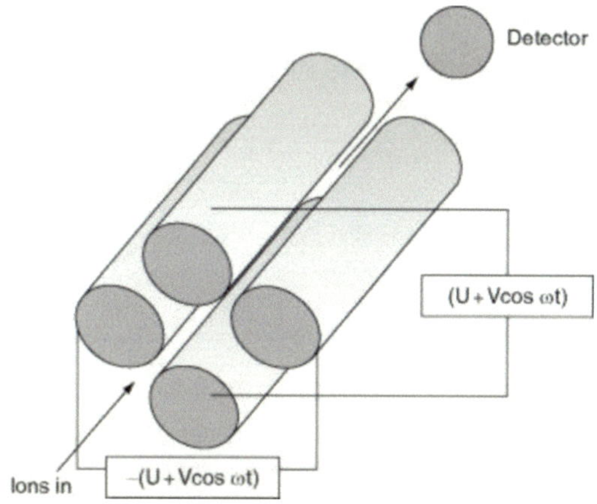

Figure 5.
Schematic diagram of operation of quadrupole mass analyzer.

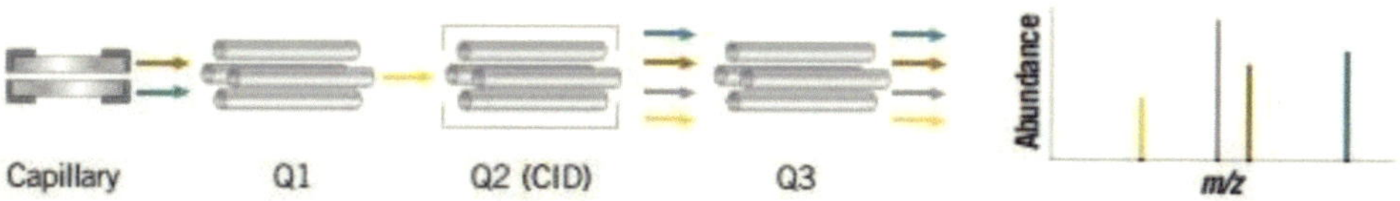

Figure 6.
Schematic diagram of triple quadrupole mass analyzer.

Based on the modes in which the mass analyzers are operated and the analytical requirements, they can be briefly classified as below:

While full scan modes are useful in understanding the total pool of masses present in the sample analyzed, product ion scan helps in obtaining structural information of a precursor ion. Precursor ion scan is suited to find structural homologs of a selected fragment ion. In multiple reaction monitoring mode (MRM), a selected parent ion (Q1 mass) is fragmented within the collision cell and selected fragment ion analyzed by the detector. Together this series of events forms a reaction where multiple ions are monitored, hence the term multiple reaction monitoring.

3.3 Linear ion trap mass analyzer

The quadrupole ion trap and the related quadrupole mass filter were invented by Paul and Steinwedel [47]. A quadrupole ion trap (QIT or 3D-IT) mass spectrometer operates with a three-dimensional quadrupole field. The QIT is formed by three electrodes: a ring electrode with a donut shape placed symmetrically between two end cap electrodes. QIT is a RF only quadrupole that acts a storage device and ions are focused to center of trap by collision with helium gas. Motion of ions in trap is regulated by axial and radial frequencies. The quadrupole ion trap can store only a limited number of ions before space charging occurs. To circumvent this effect, most instruments have an automatic gain control procedure (AGC). This procedure exactly determines the adequate fill time of the trap to maximize sensitivity and minimize resolution losses due to space charge. Ion motion can be modified either by exciting the radial or the axial frequencies by applying a small oscillating potential at the end cap electrodes during the RF ramp. Linear ion trap enables higher sensitivity than triple quadrupole mass spec analyzers in full scan mode, given the capability of ion accumulation before traveling to the detector (**Figure 7**).

Mode	Q0	Q1	Q2	Q3	Analysis type
Full scan mode (Q1 MS)	RF only	RF/ DC	RF only	RF only	Qualitative
Full scan mode (Q3 MS)	RF only	RF only	RF only	RF/ DC	Qualitative
Product ion scan (PI)	RF only	RF/ DC	RF only	RF only	Qualitative
Precursor ion scan (PC)	RF only	RF only	RF only	RF/ DC	Qualitative
Multiple reaction monitoring (MRM)	RF only	RF/ DC	RF only	RF/ DC	Quantitative

Table 1.
Modes of triple quadrupole operation and analytical requirements.

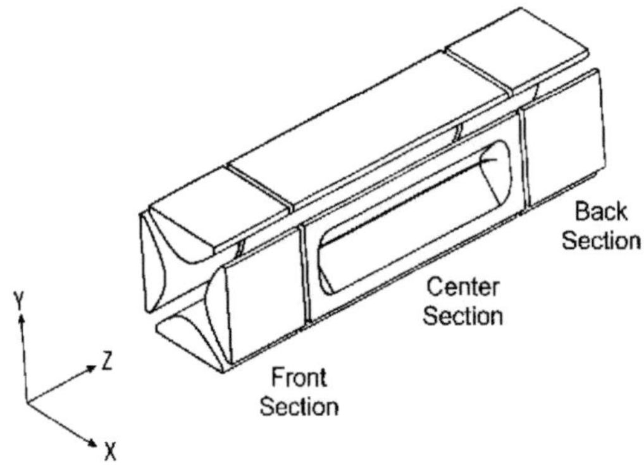

Figure 7.
Schematic diagram of linear ion trap mass analyzer.

There are more than a few important features which impact the time necessary to attain a mass spectrum (duty cycle): (i) injection time (0.5–500.0 ms), (ii) scan speed (5000–20,000 m/z units/s), (iii) separation of the parent ion and fragmentation in tandem MS or MS^n. Contrarily to the triple quadrupole, MS/MS is not performed in space but in time. Fragmentation happens with the assistance of helium as collision gas. Also, duty cycles for fragmentation (MS/MS) are much shorter in linear ion trap when compared to triple quadrupole mass analyzer. One major challenge in linear ion trap is to trap precursor ion and fragment in the same space. Often, due to this disadvantage, fragmentation spectra generated in linear ion trap differs from that of triple quadrupole CID. Also, number of MRM transitions that can be monitored in linear ion trap are quite less [4–8] when compared to QqQ mass analyzers (~100 MRM transitions can be monitored).

Due to the high sensitivity in MS^n mode, ion traps are particularly attractive for qualitative analysis in drug metabolism, metabolomics and proteomics studies. Similar sensitivities to QqQ mass analyzer can be achieved for quantitative analysis on linear ion trap, but at the price of precision and accuracy.

While linear ion traps mainly function on radial ejection, next generation mass analyzers called quadrupole linear ion trap use axial ejection. This led to discovery of hybrid triple quadrupole mass analyzers, where Q3 performs the function of both quadrupole and linear ion trap [48, 49]. Unlike linear ion trap that fragments precursor in time, these hybrid analyzers perform fragmentation in space.

The major advantage of this analyzer is that qualitative and quantitative analysis can be performed in the same LC–MS run.

3.4 Time of flight mass analyzer

Discovered in 1940's, time of flight mass analyzers achieved popularity after 1990's. Time of flight operates on principle of "time that ions need to cross in a field free tube of about 1 m length" [50, 51]. The motion of an ion is characterized by its kinetic energy Ec = 0.5 m x v^2 (m = mass, v = speed). Therefore, the time ions fly through the tube is directly proportional to their m/z value. The velocity of the ions formed is generally low and they are accelerated by strong electric fields (2000–2030,000 V) in the direction of the detector. Low mass ions reach the detector more rapidly than high mass ions. Due to the short flight time (50–100 msec) and the good transmission, a spectrum can be generated within 100 ms over an almost unlimited mass range. Mass resolution of time of flight mass analyzer depends on the length of flight tube and reduced kinetic energy spread of the ions. Length of flight tube is directly proportional to mass resolution. Kinetic energy spread can be reduced by increasing time delay between ion formation and acceleration, also known as delayed pulse extraction. Also, positioning of electrostatic mirror in the drift region of ions increases the mass resolution (**Figure 8**).

Briefly, the ions with high energy penetrate deeper into the ion mirror region than those with the same m/z at a lower energy. Because of the different trajectories, all ions of the same m/z reach the detector at the same time. With the reflectron the flight path is increased without changing the physical size of the instrument. Commercial TOF instruments are available to operate in either linear mode or reflectron mode. Even though ESI can be coupled with TOF, but the combination of MALDI and TOF is most popular as both operate on the principle of pulsed technique. Coupling of ESI with TOF needs orthogonal acceleration to drive continuous beam of ions [52].

Time of flight instruments are designed to use for qualitative analysis with MALDI or atmospheric pressure ionization. MALDI hyphenated with time of flight analyzer enables the identification of large molecules such as proteins, peptides, lipids and polymers. MS/MS information can also be obtained by CID in drift tube with the assistance of nitrogen or argon as collision gas. However, as quadrupole

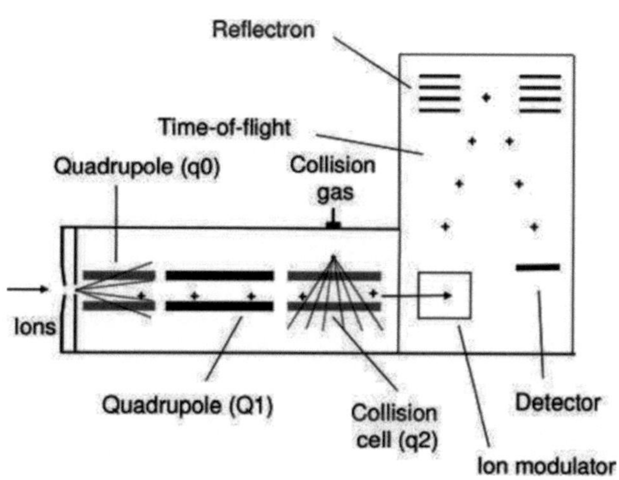

Figure 8.
Schematics representation of a quadrupole time of flight mass spectrometer.

technology is so successful for both qualitative and quantitative analysis, TOF analyzers are used as a hybrid platform with quadrupole analyzers. In these hybrid systems, TOF analyzer replaces Q3 of a triple quadrupole system. These hybrid systems are termed as QTOF mass spectrometers. QTOF systems offer high mass resolution (~40,000) and sensitivity. Accurate mass measurements are especially useful in metabolite identification studies and peptide analysis [53]. Various other hydrid TOF platforms have been reported including, linear ion trap, quadrupole ion trap and TOF-TOF mass spectrometers [54–56].

3.5 Orbitrap mass analyzer

Orbitrap mass analyzer operates on principle of Fourier transform, where orbital trapping of ions around an electrode system is achieved with the assistance of electrical field [57]. The orbitrap is formed by a central spindle-like electrode surrounded by an electrode with a barrel-like shape to create an electrostatic potential. The m/z is a reciprocal proportionate to the frequency of the ions oscillating along the z-axis. Detection is performed by measuring the current image of the axial motion of the ions around the inner electrode. The mass spectrum is obtained after Fourier transformation of the image current. The orbitrap provides a mass resolving power exceeding 100,000 and a mass accuracy ~3 ppm. To be operational as a mass spectrometer the orbitrap requires external ion accumulation, cooling and fragmentation (**Figure 9**).

The first commercial instrument to utilize this capability, LTQ Orbitrap Classic, was introduced by Thermo Fisher Scientific in 2005, which later underwent many innovations with the addition of a collision cell after the C-trap in LTQ Orbitrap XL, addition of electron transfer dissociation (ETD) capabilities, followed by MALDI source operating at reduced pressure with high-end LTQ Orbitrap XL MALDI instrument, and finally a stacked ring rf ion guide (so called S-lens) brought about 10-fold higher transfer efficiency in the MS/MS mode.

Typically, the highest resolving powers available in TOF devices are several times lower than the resolution in Orbitrap, although recent multipass TOF devices

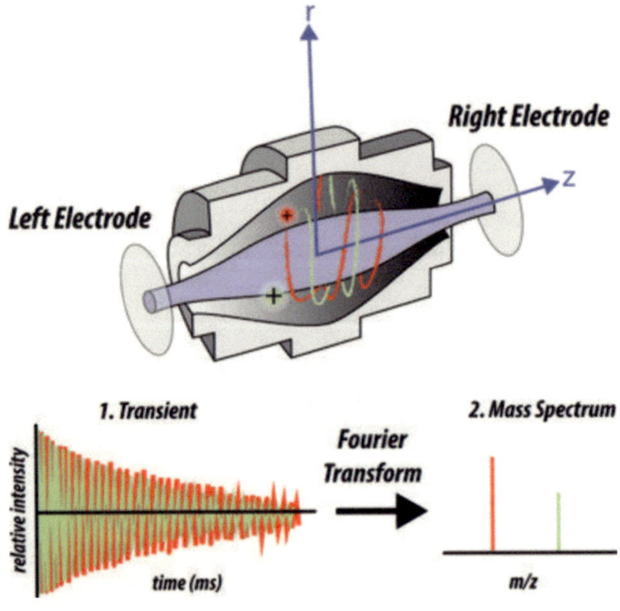

Figure 9.
Schematic representation of operation of orbitrap mass analyzer.

are capable of ultrahigh resolution (R ≥ 100,000 at m/z 400) [58, 59]. While TOF accompanies similar resolution in both MS and MS/MS modes, orbitrap suffers from low resolution in MS/MS mode. Orbitrap technology will endure to progress towards increased resolving power, acquisition speed, sensitivity and mass accuracy. These developments will indisputably open the arena for new applications as the Orbitrap instruments are getting more prevalent and exploring into new areas of research.

4. Applications in drug discovery

4.1 ADME studies

Drug discovery research was solely driven by chemists and pharmacologist in early 1990's, when very little is known about drug absorption, distribution, metabolism and elimination (ADME). However, it did not take much time before researchers realized the importance of optimizing ADME properties of NCE's for successfully driving drug discovery programs [60]. In this section, we highlighted importance of ADME in drug discovery and its relation to mass spectrometry.

Drug metabolism also known as xenobiotic biotransformation is the process by which lipophilic compounds gets eliminated from the body after getting converted to hydrophilic species that are easily filtered through kidney. While metabolism is desired in few cases where metabolites are the active species producing efficacy, there are metabolites that are toxic in nature. In such cases, where toxic by-products are produced, metabolism is not desired. Metabolism as a discipline drawn its first attention after the publication of RT Williams on Detoxification mechanisms [61].

Drug metabolism over the years with the aid of mass spectrometry technology has evolved in understanding the metabolic pathways of NCE's and also to identify the metabolites (both desired and undesired) [62–65]. Mass spectrometry was initially hyphenated with gas chromatography to understand the metabolic behavior of NCE's. Gas chromatography worked well for analyzing volatile compounds and its metabolites, however it did not suit for nonvolatile and thermolabile compounds. With the advent of liquid chromatography that can handle and separate components without subjecting to evaporation, it became prevalent as an analytical tool for understanding drug metabolism in drug discovery and development [66, 67].

As a part of understanding the metabolic properties, NCE's will be initially screened for metabolic stability in across species (human/rat/dog/mouse/monkey) and in various matrices including microsomes/S9 fractions/cytosol/hepatocytes, plasma, tissue homogenates, and buffer. If metabolism is not desired then compounds will be screened for their stability in relevant matrices and compounds with moderate to high stability (defined by half-life and intrinsic clearance) are further optimized for additional ADME properties. Various Phase 1 metabolic reactions including oxidation, demethylation, hydroxylation and phase 2 metabolic reactions covering glucuronidation, sulfation, methylation, amino acids conjugation and glutathione conjugation can be quantitatively and qualitatively studied using LC–MS/MS. Additionally, for compounds that are unstable, understanding the soft spots responsible for instability helps medicinal chemists to make relevant structural modifications in order to stabilize the unstable compounds. Understanding the soft spots precisely, needs the assistance of high-resolution mass spectrometry instruments such as TOF and Orbitrap. With the accurate mass information obtained from these mass spec's, identifying a metabolite structure will be spot on.

Similarly, other in vitro parameters such as permeability, protein binding, solubility, lipophilicity, CYP inhibition and CYP induction also play a key role in drug disposition [15, 68, 69]. All of these assays have high sensitivity requirements and demand quantification of analytes within a few nanomolar range. For example, in case of permeability assessment, low permeable compounds such as atenolol permeate poorly from apical to basolateral side complicates the quantification of the apparent permeability values, if analyzed with low sensitive detectors. Likewise, fraction unbound values for highly protein compounds such as warfarin were such low that it demands highly sensitive detectors to accurately quantify such low levels. Needless to mention that all of the assays performed to optimize ADME properties of NCE's require highly sensitive detection systems. Hence, with its superior detection sensitivity, mass spectrometer has become an indispensable tool to understand the in vitro ADME properties of NCE's.

As discussed in the previous sections, even though there exist many ionization techniques, atmospheric pressure ionization (API) was more successful in drug metabolism studies, given its rapid, specific and sensitive methodologies for the identification of drugs and its metabolites. Mass spectrometer instruments types used in ADME studies vary from those that provide nominal mass information and accurate mass information. Nominal mass instruments such as triple quadrupoles are useful for quantitative applications, whereas accurate mass instruments including QTOF and Orbitrap are used for both quantitative and qualitative applications.

Of the atmospheric pressure ionization techniques available, ESI and APCI are the most commonly used. While APCI can accommodate high flow rates and produce high sensitivity, nevertheless analytes are subjected to higher temperatures in the evaporation process and hence as an ionization technique is not suitable for thermolabile (esp. glucuronides, N-oxides and sulfates) compounds. However, ESI is comparatively a soft ionization technique and could efficiently ionize these fragile compounds without degradation. As a whole, mass spectrometer exhibits both qualitative and quantitative applications in drug metabolism studies. However, in case of other ADME battery of assays as described above, mass spectrometer is used majorly for quantitative applications.

4.2 Metabolite identification

Metabolite identification (Met-ID) provides a variety of inputs in drug discovery and development which includes in vitro metabolite profiling in early stages of lead identification/optimization, followed by in vitro/in vivo correlation in late stage lead identification, characterization of putative metabolites, cross-species comparison to identify the right tox preclinical species, understanding drug–drug interactions, and identifying pharmacologically active or toxic metabolites and the mechanisms by which they are formed [70, 71].

Met-ID is quite challenging when it comes to a) identification of vast number of diverse metabolites, b) metabolites that are of low abundance, and c) high throughput analytical requirements to screen majority of early leads in preclinical drug discovery [72]. Even though there exist various platforms such as triple quadrupole, linear ion trap, and Qtrap to quantitatively/qualitatively identify metabolites, they turned obsolete due to nominal mass information they generate and are no longer valuable [73].

The term mass resolution is used to describe the mass resolving power according to the degree to which two analytes with close m/z values can be separated and identified. A practical and convenient way of evaluating the mass resolution of an instrument is the use of the full width at half maximum (FWHM) definition in which the $m/z/\Delta m/z$ ratio is calculated, where m/z is the mass-to-charge value of an ion peak and $\Delta m/z$ is the full width at half the maximal height of the peak. Nominal mass instruments generate resolution in low thousands (1000–4000), which cannot separate isobaric ions with similar nominal m/z value. However, high resolution mass spectrometers provide resolution in the range of higher thousands (10000–100,000), that successfully identify and separate isobaric ions (**Figure 10**).

For a given nominal mass, there exists many possible molecular structures for an assigned biotransformation pathway. Hence, for an accurate molecular structure the metabolite has to be scaled up in larger quantities and measured using NMR spectroscopy. However, scaling up metabolites to "mg" quantities need tremendous efforts, resources and is not an economical approach. Later, with the discovery of high-resolution mass spectrometers such as TOF and Orbitrap, that provides accurate mass to the fourth decimal, enabled accurate prediction of molecular structures of metabolites [74–79].

In principle, as long as data can be measured accurately, high-resolution data is sufficient to demonstrate the presence or absence of defined species. Apart from high resolution masses, modern mass spectrometers also generate data with higher accuracy. The term mass accuracy is used to define how close the mass measured by the mass spectrometer is to the theoretical exact mass of an ion. Mass accuracy

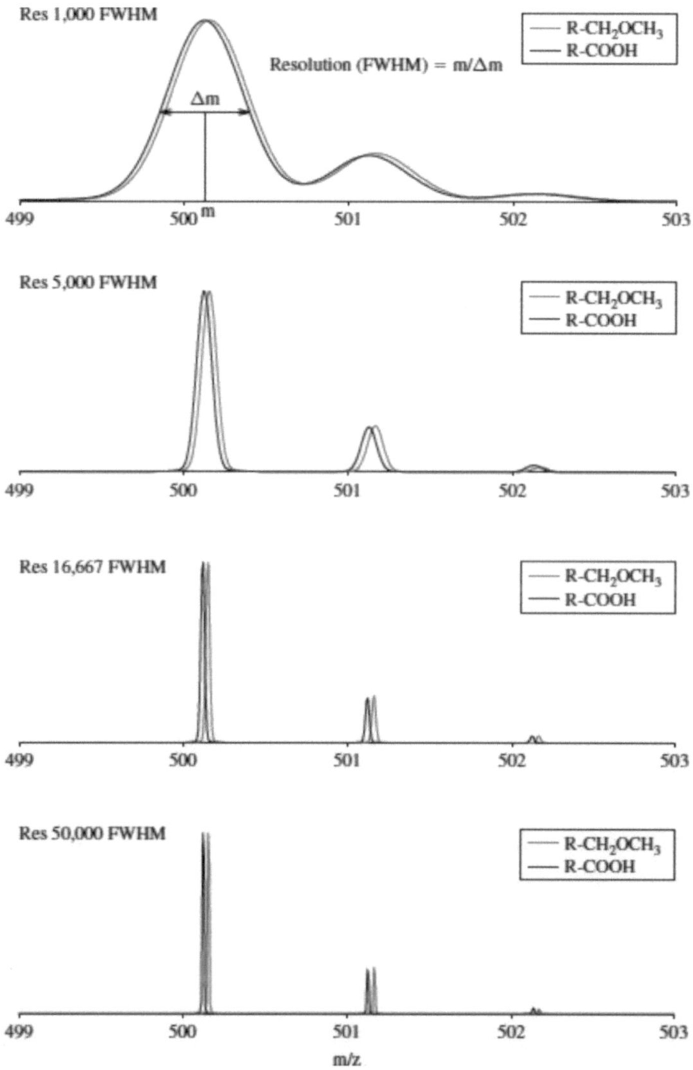

Figure 10.
Increased resolution separating two closely arranged analytes with similar nominal m/z values.

is typically expressed as a relative mass error using the ratio of the difference between the experimental and theoretical m/z values over the theoretical m/z value of an ion.

Metabolite identification studies are typically performed in full scan mode (or) data dependent scan mode. In full scan mode, accurate mass information of parent ions is studied to understand the biotransformation pathways. However, to propose soft spots for metabolites, data dependent scans consisting of combination of full scan and product ion scans are performed. Metabolites in general can be considered as off springs to the parent molecular ions that carry similar fragments as that of parent ion (or) neutral loss fragments. In case of similar fragments to that of parent ion, these fragments can be considered unmodified and are similar to that of parent. Whereas, if the fragment ions are accompanied by a mass change for a given biotransformation pathway, then metabolite soft spot can be proposed with precision to that fragment ion (for example, in case of hydroxylation, neutral loss fragment in metabolite carries an additional mass of 16 amu) [80]. Additionally, apart from fragmentation scans accompanied with full scan, few other specific dependent scans consisting of neutral loss scan, and precursor ion scans are used to study various biotransformation pathways including glucuronidation, glutathione conjugation, sulfation (for example, glucuronidation is accompanied by a specific neutral loss of 176 da and glutathione conjugation by a neutral loss of 129 da in positive mode) [81, 82]. These specific scan functions are helpful in eliminating the background noise and identify the metabolites that exist even at lower abundance.

One major challenge in metabolite identification using LC–MS technology, is quantifying the relative abundance of metabolites. As mass spec quantitation is accompanied by many source/gas and compound dependent parameters that aid in the efficient ionization and detection, minor modifications in the metabolite structures alter the sensitivity by few orders of magnitude. Hence, quantitative results using mass spectrometer for the metabolites for which synthetic standards are not available, is not feasible. Additionally, it is difficult to synthesize each and every metabolite and determine their concentrations accurately. Alternatively, few researchers used LC-UV hyphenated with mass spectrometer to measure the relative abundance of metabolites. However, as majority of metabolites exist in low abundance, it becomes difficult to measure their relative abundance by UV spectroscopy. Also, UV is prone to differences in analytical sensitivities with minor modifications in structure. Hence, future mass spectrometers need to be designed to address these key concerns and facilitate evaluation of both quantitative and qualitative aspects of metabolites.

4.3 Pharmacokinetic analysis

Screening paradigm in drug discovery includes evaluation of compounds for their ADME properties by various in vitro assays. The datasets obtained from these assays help in rationalizing the synthetic chemistry efforts and make progress towards a pool of lead compounds that exhibit promising in vitro ADME properties. However, in vitro screening consists of unique assay platforms that can only answer a specific question of interest and can never cover all aspects of complex biological systems. Hence, it warrants the screening of selected lead compounds in preclinical species such as rats, mice, dog, pig and monkey, before progressing to clinic [83, 84]. Role of LC–MS/MS in performing bioanalysis of pharmacokinetic samples was well reported in the literature [85–88].

Design of pharmacokinetic studies varies from single route administration with few sampling points followed by multiple route administration with detailed time course evaluation. Samples of various natures ranging from blood, plasma, serum are collected to analyze the systemic concentration levels. Additionally, to understand the tissue distribution of compounds, various tissues including liver, intestine, brain, spinal cord, heart, lungs, kidneys, skin and adipose tissue are also analyzed. Initially, lead compounds are dosed intravenously to understand the disposition parameters such as volume of distribution, clearance, half-life and mean residence time. Compounds that possess decent pharmacokinetic parameters in intravenous route are further evaluated in alternate (enteral and parenteral) routes to assess bioavailability and exposure parameters (Cmax, Tmax, $AUC_{0\text{-last}}$). To determine plasma concentrations as low as "ng" levels, sensitive analytical detectors are needed. HPLC-UV detector systems are proven robust for analyzing concentrations at higher "μg" level. However, as UV detection is not specific, it suffers from high background noise when biological samples are analyzed. High background noise in turn causes quantitation issues at the lower portion of calibration curve. On the other hand, LC–MS/MS detection system is considered to be highly specific and selective, as background noise can be eliminated by analyzing selected analytes of interest with desired m/z ratios. Also, mass spectrometric detection is considered highly sensitive than UV spectroscopic detection. Altogether, these advanced features, enabled LC–MS/MS to overcome the limitations of UV detection and is more frequently used in drug discovery and development for the bioanalysis of pharmacokinetic samples.

Of the various mass spectrometers available in the market, triple quadrupole LC–MS/MS systems have demonstrated tremendous success when it comes to quantitative applications. Monitoring specific reaction transitions that consist of parent and fragment ion, with associated source/gas and compound dependent parameters makes the mass spectrometer highly specific, selective and sensitive. Pharmacokinetic study samples are in general monitored for plasma concentrations over the time profile and hence triple quadrupole systems have achieved greater success. However, exploratory studies performed to understand metabolic pattern and biotransformation mechanisms of NCE's again need the assistance of linear ion traps and high-resolution mass spectrometer platforms (Orbitrap and TOF). As pharmacokinetic study samples are of biological origin and complex in nature, extracts obtained after sample preparation complicate the bioanalysis on LC–MS/MS. This phenomenon in broad terms is termed as Matrix effects. Causes of matrix effects and strategies to mitigate these effects are discussed in detail in the succeeding section.

Typically, pharmacokinetic studies are performed by administering a single test item in preclinical species. However, screening of single test item is labor intensive, not economical and demand higher turn-around times to generate pharmacokinetic data. Hence, researchers have come up with an alternate strategy, where a pool of compounds are administered in single dose, a technique well-known as cassette dosing or N-in-one dosing. The foundation for designing cassette dosing strategy comes from the ability of LC–MS/MS to analyze multiple test items without any chromatographic separation [89–92]. However, disadvantages with cassette dosing include altered pharmacokinetics due to drug–drug interaction potential, non-feasibility of pooling compounds with close molecular weights and compounds with differing physicochemical properties posing formulatability issues. Compounds with differing physicochemical properties comes with a challenge of formulating the selected pool of compounds in a single formulation vehicle. While issues with physicochemical properties and close molecular weights can be taken care of, drug–drug interaction appears to be of a major concern. One approach to minimize DDI's is by administering the compounds at minimal doses, collectively not exceeding the dose of single test item administration [93]. Also, dosing volumes can be kept as low as possible. Additionally, along with pool of unknown compounds, a quality control compound with known pharmacokinetic parameters can be administered. Pharmacokinetic study results of unknown compounds can be considered acceptable, as long as quality control compounds fall within the set acceptance criteria. Compounds with less than 5 da difference in molecular weight are difficult to pool, when triple quadrupole systems are used for analyzing the pharmacokinetic samples. However, this challenge can be overcome by considering high resolution mass spectrometric analysis. Typical pharmacokinetic parameters studied from intravenous administration include half-life ($t_{1/2}$), clearance (Cl), volume of distribution (V_z), mean residence time (MRT), area under the curve ($AUC_{0\text{-last}}$; $AUC_{0\text{-inf}}$), whereas parameters such as Cmax, Tmax and area under the curve are studies in other routes of administration. When compounds are dosed in multiple routes, along with intravenous route of administration, absolute bioavailability values are calculated.

Current fast LC–MS/MS instruments enable analysis of mixture of analytes with minimal separation, shorter run times and also feasible for hyphenation with ultra-fast liquid chromatography systems. Additionally, latest LC–MS/MS systems are capable of analyzing thousands of samples every week due to higher loading capacity of samples in autosampler, shorter run times and introduction of 96/384 well plate formats.

Test samples are analyzed against a calibration curve and a set of quality control samples. Calibration curve consists of 8–10 known standards and quality controls span the calibration curve at a minimum of 3 levels. Typical accuracy limits for qualifying the calibration and quality control samples is set as ±20%. Typical turnaround times for execution of pharmacokinetic studies right from dosing initiation to generation of pharmacokinetic parameters spans 1–2 weeks. Importance of high throughput bioanalysis and its role in drug discovery is discussed in detail in the section below.

4.4 High throughput bioanalysis

With the combinatorial chemistry efforts leading to synthesis of hundreds of compounds in early phases of drug discovery, there is a constant need for analytical platforms that can quickly churn out data and help in accelerating the discovery process. LC–MS/MS with proven track record as a reliable analytical platform had undergone evolutionary changes to support high throughput demands of drug discovery [60, 70, 87, 94–96]. These strategies include advancements in chromatographic columns, where lower dimensions and microbore HPLC columns cut short run time to one minute per sample. Quick sample run times help in analyzing higher number of samples with in the given stipulated time [97]. Faster gradient methods with LC pumps that can handle higher pressure also enable analysis with shorter run times [98]. With the invention of monolithic HPLC columns that can be operated at high flow rates in the order of 5–6 mL/min, analytical run times were significantly reduced [99–101]. One disadvantage of these columns is that the high flow rates translate in to higher usage of mobile phase, making it an expensive alternative.

Another approach for increasing sample throughput is by the use of parallel HPLC columns, where the effluent from two HPLC systems could be combined and assayed by using the MRM/SRM capabilities of the MS/MS system [102, 103]. One more approach in enhancing the throughput is through staggered analysis approach. Here, multiple HPLC columns are used, but the injection time is staggered such that the “analytical window” can be selected sequentially in order to maximize the use of the MS/MS system and increase sample throughput [104].

On the other hand, throughput can also be increased by pooling of samples, provided the sensitivity is not seriously compromised. When these samples are analyzed on triple quadrupole systems, care must be taken to have molecular weight differences by at least 5 da. Typically, 4–5 analytes can be pooled and analyzed. Only challenge with pooling/cassette strategy (even with high resolution mass spectrometers) is its non-suitability for isomeric compounds. With advancements in instrumentation technology, modern LC autosamplers are designed to accommodate higher sample load. There are autosamplers that can accommodate as high as twelve 96 well plates. Higher loading capacity of autosamplers enable unattended analysis of large number of samples (**Figure 11**).

When it comes to sample preparation, robotic platforms can be used to screen larger pool of compounds across various in vitro assays. However, this strategy can also be used for processing of in vivo samples, provided if the sample cohort is higher. In general, automated robotic sample preparation platforms are quite often used for screening of compounds in in vitro assays. These robotic platforms help in decreasing the manpower involved and time taken for performing the assays. However, main disadvantage with robotic platforms is the need for preparation of larger volumes of reagents and the cost factor involved. Hence, it is not recommended to use these platforms unless there exists a larger library of compounds.

4.5 Matrix effects

With the sample nature being biological in origin, supernatants obtained after sample preparation consists of many endogenous components that compete with the analyte of interest and result in either suppression/enhancement of ionization. This process of alteration of the ionization of analytes is termed as matrix effects. The "matrix" refers to all components in the sample other than analyte(s) of interest. Matrix effects are defined as "interference from matrix components that are

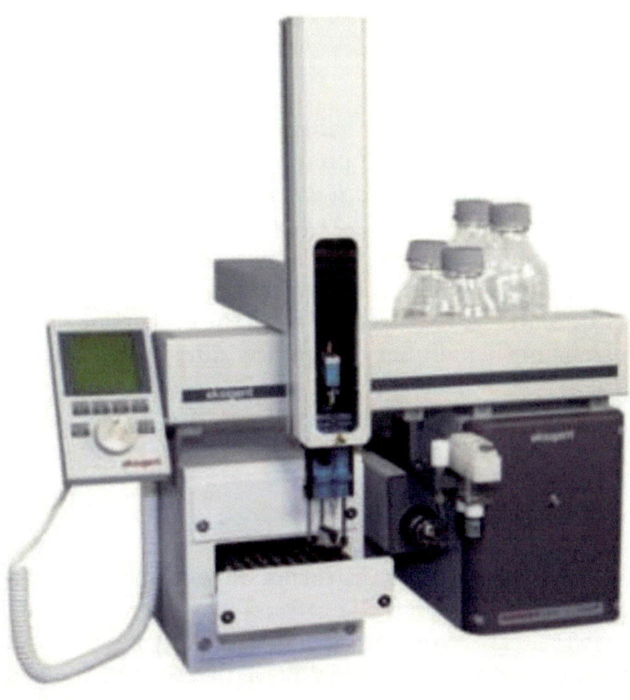

Figure 11.
Pictorial representation of LC–MS/MS system with higher loading capacity of autosampler.

unrelated to the analyte" [105, 106]. The process of ion suppression/enhancement is in general referred as matrix effect and is main subject of various published reviews [107–112].Matrix effects result in significant deviation in precision and accuracy of results which in turn debate the reliability of pharmacokinetic parameters of NCE's generated. Matrix effect alters the sensitivity, reproducibility and challenges the reliability of analytical techniques. Although matrix effects occur as a result of various exogenous and endogenous components, one major area of concern is formulation excipients (an exogenous component) used in the preparation of formulations. Dosing vehicles are generally used at high concentrations to solubilize test articles of highly lipophilic nature [113–115]. This in turn can be instrumental in causing matrix effects, thereby questioning the reliability of preclinical PK parameters. This phenomenon has been reported by us in the past for various excipients such as PEG 400 [110, 116, 117], Cremophor EL [111, 118] and Solutol HS15 [112, 119].

Several mechanisms have been proposed to explain matrix effects, but the exact process remains uncertain [120, 121]. Various mechanisms by which matrix components cause ion suppression are as follows:

- Charge competition between analyte and matrix components [122, 123].
- Change in droplet surface tension leading to formation of large droplets and insufficient desolvation [121, 124].
- Preferential ion evaporation due to matrix components gathering at droplet surface.
- Change in mass of analyte ion due to ion pairing and adduct formation
- Co-precipitation with non-volatile matrix components [125].
- Gas phase deprotonation.

Reduction of matrix effects can be achieved through various strategies including decreasing the level of matrix components, improving chromatographic separation of interfering materials from the analyte, various sample preparation strategies, lower injection volumes, and even by simple dilution of samples to reduce the overall concentrations of both analyte and co-extracted materials [126, 127]. Switching ionization sources will also help in mitigating the matrix effects [112, 116, 118, 119]. Matrix effects occurring in the early time point samples can be monitored, using another aliquot of the early time point samples analyzed at a higher dilution [128].

5. Conclusion

Over the past few decades, technological advancements in mass spectrometer enabled it to surpass other detection platforms and evolve as an indispensable analytical tool to support the bioanalytical needs of drug discovery and development. Current generation mass spectrometers could efficiently handle both qualitative and quantitative aspects of bioanalysis. Additionally, the likelihood of hyphenation of mass spectrometers with ultrafast liquid chromatography systems, extended its applications to high throughput bioanalysis. Even though significant achievements were made in the past, instruments will continue to get more and more sensitive and become better acquiescent to automation.

Author details

Vijayabhaskar Veeravalli[1*], Lakshmi Mohan Vamsi Madgula[1]
and Pratima Srivastava[2]

1 Syngene International Limited, Biocon Park, SEZ, Bangalore India

2 GVK BIO Pvt Ltd, Hyderabad, Telangana, India

*Address all correspondence to: vveerav1@jhmi.edu

References

[1] Thomson JJ. Rays of positive electricity and their application to chemical analyses. Proceedings of the Royal Society. 1913;**89**:1-20

[2] Gohlke RS. Time-of-flight mass spectrometry and gas–liquid partition chromatography. Analytical Chemistry. 1959;**31**:535-541

[3] Arpino P, Baldwin MA, McLafferty FW. Liquid chromatography–mass spectrometry. II. Continuous monitoring. Biomedical Mass Spectrometry. 1974;**1**:80-82

[4] Ito Y, Takeuchi T, Ishii D, Goto M. Direct coupling of micro high performance liquid chromatography with fast atom bombardment mass spectrometry. Journal of Chromatography. 1985;**346**:161-166

[5] Caprioli RM, Fan T, Cottrell JS. A continuous-flow sample probe for fast atom bombardment mass spectrometry. Analytical Chemistry. 1986;**58**:2949-2954

[6] Blakley CR, Vestal ML. Thermospray interface for liquid chromatography-mass spectrometry. Analytical Chemistry. 1983;**55**:750-754

[7] Willoughby RC, Browner RF. Monodisperse aerosol generation interface for combining liquid chromatography with mass spectroscopy. Analytical Chemistry. 1984;**56**:2626-2631

[8] Ho CS, Lam CW, Chan MH, Cheung RC, Law LK, Lit LC, et al. Electrospray ionisation mass spectrometry: Principles and clinical applications. Clinical Biochemist Reviews. 2003;**24**(1):3-12

[9] Frank R, Hargreaves R. Clinical biomarkers in drug discovery and development. Nature Reviews. Drug Discovery. 2003;**2**(7):566-580

[10] Lee MS. Integrated Strategies for Drug Discovery Using Mass Spectrometry. Vol. 550. Hoboken: John Wiley & Sons; 2005

[11] Korfmacher W. Using Mass Spectrometry for Drug Metabolism Studies. Vol. 370. Boca Raton: CRC Press; 2005

[12] Korfmacher WA. Lead optimization strategies as part of a drug metabolism environment. Current Opinion in Drug Discovery & Development. 2003;**6**(4):481-485

[13] Korfmacher WA. Principles and applications of LC-MS in new drug discovery. Drug Discovery Today. 2005;**10**(20):1357-1367

[14] Roberts SA. High-throughput screening approaches for investigating drug metabolism and pharmacokinetics. Xenobiotica. 2001;**31**(8-9):557-589

[15] Eddershaw PJ, Beresford AP, Bayliss MK. ADME/PK as part of a rational approach to drug discovery. Drug Discovery Today. 2000;**5**(9):409-414

[16] Kumar GN, Surapaneni S. Role of drug metabolism in drug discovery and development. Medicinal Research Reviews. 2001;**21**(5):397-411

[17] Riley RJ, Martin IJ, Cooper AE. The influence of DMPK as an integrated partner in modern drug discovery. Current Drug Metabolism. 2002;**3**(5):527-550

[18] Beine B, Diehl HC, Meyer HE, Henkel C. Tissue MALDI mass spectrometry imaging (MALDI MSI) of peptides. Methods in Molecular Biology. 2016;**1394**:129-150

[19] Takats Z, Wiseman JM, Gologan B, Cooks RG. Mass spectrometry sampling under ambient conditions with desorption electrospray ionization. Science. 2004;**306**(5695):471-473

[20] Cody RB, Laramee JA, Durst HD. Versatile new ion source for the analysis of materials in open air under ambient conditions. Analytical Chemistry. 2005;**77**(8):2297-2302

[21] Liigand P, Kaupmees K, Haav K, Liigand J, Leito I, Girod M, et al. Think negative: Finding the best electrospray ionization/MS mode for your Analyte. Analytical Chemistry. 2017;**89**(11):5665-5668

[22] Lubin A, Bajic S, Cabooter D, Augustijns P, Cuyckens F. Atmospheric pressure ionization using a high voltage target compared to electrospray ionization. Journal of the American Society for Mass Spectrometry. 2017;**28**(2):286-293

[23] Wilm M. Principles of electrospray ionization. Molecular & Cellular Poroteomics. 2011;**10**(7):M111.009407

[24] Pozniak BP, Cole RB. Perspective on electrospray ionization and its relation to electrochemistry. Journal of the American Society for Mass Spectrometry. 2015;**26**(3):369-385

[25] Carroll DI, Dzidic I, Stillwell RN, Haegele KD, Horning EC. Atmospheric pressure ionization mass spectrometry. Corona discharge ion source for use in a liquid chromatograph-mass spectrometer computer analytical system. Analytical Chemistry. 1975;**47**:2369-2372

[26] Beccaria M, Inferrera V, Rigano F, Gorynski K, Purcaro G, Pawliszyn J, et al. Highly informative multiclass profiling of lipids by ultra-high performance liquid chromatography - low resolution (quadrupole) mass spectrometry by using electrospray ionization and atmospheric pressure chemical ionization interfaces. Journal of Chromatography. A. 2017;**1509**:69-82

[27] Galani JHY, Houbraken M, Van Hulle M, Spanoghe P. Comparison of electrospray and UniSpray, a novel atmospheric pressure ionization interface, for LC-MS/MS analysis of 81 pesticide residues in food and water matrices. Analytical and Bioanalytical Chemistry. 2019;**411**:5099-5113

[28] Robb DB, Covey TR, Bruins AP. Atmospheric pressure photoionisation: An ionization method for liquid chromatography-mass spectrometry. Analytical Chemistry. 2000;**72**(15):3653-3659

[29] Raffaelli A, Saba A. Atmospheric pressure photoionization mass spectrometry. Mass Spectrometry Reviews. 2003;**22**(5):318-331

[30] Bos SJ, van Leeuwen SM, Karst U. From fundamentals to applications: Recent developments in atmospheric pressure photoionization mass spectrometry. Analytical and Bioanalytical Chemistry. 2006;**384**(1):85-99

[31] Grun CH, Besseau S. Normal-phase liquid chromatography-atmospheric-pressure photoionization-mass spectrometry analysis of cholesterol and phytosterol oxidation products. Journal of Chromatography. A. 2016;**1439**:74-81

[32] Kauppila TJ, Syage JA, Benter T. Recent developments in atmospheric pressure photoionization-mass spectrometry. Mass Spectrometry Reviews. 2017;**36**(3):423-449

[33] Vaikkinen A, Kauppila TJ, Kostiainen R. Charge exchange reaction in dopant-assisted atmospheric pressure chemical ionization and atmospheric pressure photoionization. Journal

of the American Society for Mass Spectrometry. 2016;**27**(8):1291-1300

[34] Karas M, Hillenkamp F. Laser desorption ionization of proteins with molecular masses exceeding 10,000 daltons. Analytical Chemistry. 1988;**60**(20):2299-2301

[35] Rappsilber J, Moniatte M, Nielsen ML, Podtelejnikov AV, Mann M. Experiences and perspectives of MALDI MS and MS/MS in proteomic research. International Journal of Mass Spectrometry. 2003;**226**(1):223-237

[36] Nielen MWF. Maldi time-of-flight mass spectrometry of synthetic polymers. Mass Spectrometry Reviews. 1999;**18**(5):309-344

[37] Gut IG. DNA analysis by MALDITOF mass spectrometry. Human Mutation. 2004;**23**:437-441

[38] Schiller J, Suss R, Arnhold J, Fuchs B, Lessig J, Muller M, et al. Matrixassisted laser desorption and ionization time-of-flight (MALDITOF) mass spectrometry in lipid and phospholipid research. Progress in Lipid Research. 2004;**43**:449-488

[39] McCombie G, Knochenmuss R. Small-molecule MALDI using the matrix suppression effect to reduce or eliminate matrix background interferences. Analytical Chemistry. 2004;**76**(17):4990-4997

[40] Donegan M, Tomlinson AJ, Nair H, Juhasz P. Controlling matrix suppression for matrix-assisted laser desorption/ionization analysis of small molecules. Rapid Commun Mass Sp. 2004;**18**(17):1885-1888

[41] Poetzsch M, Steuer AE, Hysek CM, Liechti ME, Kraemer T. Development of a high-speed MALDI-triple quadrupole mass spectrometric method for the determination of 3,4-methylenedioxymethamphetamine (MDMA) in oral fluid. Drug Testing and Analysis. 2016;**8**(2):235-240

[42] Gobey J, Cole M, Janiszewski J, Covey T, Chau T, Kovarik P, et al. Characterization and performance of MALDI on a triple quadrupole mass spectrometer for analysis and quantification of small molecules. Analytical Chemistry. 2005;**77**(17):5643-5654

[43] March RE, Todd JFJ. Quadrupole Ion Trap Mass Spectrometry. 2nd ed. Vol. 165. Hoboken, New Jersey, USA: Wiley-Interscience; 2005. p. 346

[44] Haag AM. Mass analyzers and mass spectrometers. Adv. Exp. Med. Biol. 2016;**919**:157-169

[45] de Hoffmann E. Tandem mass spectrometry: A primer. Journal of Mass Spectrometry. 1996;**31**:129-137

[46] Schwartz JC, Wade AP, Enke CG, Cooks RG. Systematic delineation of scan modes in multidimensional mass spectrometry. Analytical Chemistry. 1990;**62**(17):1809-1818

[47] Paul W, Steinwedel H. A new mass spectrometer without magnetic field. Zeitschrift für Naturforschung. 1953;**8a**:448-450

[48] Hager JW. A new linear ion trap mass spectrometer. Rapid Communications in Mass Spectrometry. 2002;**16**:512-526

[49] Hopfgartner G, Zell M. Q trap MS: A new tool for metabolite identification. In: Using Mass Spectrometry for Drug Metabolism Studies. Boca Raton, Florida, USA: CRC Press; 2004

[50] Mamyrin BA. Time-of-flight mass spectrometry (concepts, achievements, and prospects). International Journal of Mass Spectrometry. 2001;**206**:251-266

[51] Uphoff A, Grotemeyer J. Tutorial: The secrets of time-of flight mass

spectrometry revealed. European Journal of Mass Spectrometry. 2003;**9**(3):151-164

[52] Guilhaus M, Selby D, Mlynski V. Orthogonal acceleration time-of-flight mass spectrometry. Mass Spectrometry Reviews. 2000;**19**(2):65-107

[53] Hopfgartner G, Chernushevich IV, Covey T, Plomley JB, Bonner R. Exact mass measurement of product ions for the structural elucidation of drug metabolites with a tandem quadrupole orthogonal-acceleration time-of-flight mass spectrometer. Journal of the American Society for Mass Spectrometry. 1999;**10**:1305-1314

[54] Martin RL, Brancia FL. Analysis of high mass peptides using a novel matrix-assisted laser desorption/ionisation quadrupole ion trap timeof-flight mass spectrometer. Rapid Communications in Mass Spectrometry. 2003;**17**:1358-1365

[55] Vestal ML, Campbell JM. Tandem time-of-flight mass spectrometry. Methods in Enzymology. 2005;**402**:79-108

[56] Suckau D, Resemann A, Schuerenberg M, Hufnagel P, Franzen J, Holle A. A novel MALDI LIFT-TOF/TOF mass spectrometer for proteomics. Analytical and Bioanalytical Chemistry. 2003;**376**(7):952-965

[57] Makarov A. Electrostatic axially harmonic orbital trapping: A high performance technique of mass analysis. Analytical Chemistry. 2000;**72**:1156-1162

[58] Verentchikov AN, Yavor M, Hasin YI, Gavrik MA. Multireflection planar time-of-flight mass analyzer. I: An analyzer for a parallel tandem spectrometer. Technical Physics. 2005;**50**(1):73-81

[59] Klitzke CF, Corilo YE, Siek K, Binkley J, Patrick J, Eberlin MN. Petroleomics by ultrahigh-resolution time-of-flight mass spectrometry. Energy & Fuels. 2012;**26**:5787-5794

[60] Korfmacher W. Bioanalytical assays in a drug discovery environment. In: Using Mass Spectrometry for Drug Metabolism Studies. Boca Raton: CRC Press; 2005. pp. 1-34

[61] Williams RT. Deoxication Mechanisms : The Metabolism and Detoxication of Drugs, Toxic Substances and Other Organic Compounds. Boca Raton, Florida, USA: Chapman and Hall; 1959

[62] Korfmacher WA, Cox KA, Bryant MS, Veals J, Ng K. HPLC-API/MS/MS: A powerful tool for integrating drug metabolism into the drug discovery process. Drug Discovery Today. 1997;**2**:532-537

[63] Yengi LG, Leung L, Kao J. The evolving role of drug metabolism in drug discovery and development. Pharmaceutical Research. 2007;**24**(5):842-858

[64] Chiu SH, Thompson KA, Vincent SH, Alvaro RF, Huskey SW, Stearns RA, et al. The role of drug metabolism in drug discovery: A case study in the selection of an oxytocin receptor antagonist for development. Toxicologic Pathology. 1995;**23**(2):124-130

[65] Oliveira EJ, Watson DG. Liquid chromatography–mass spectrometry in the study of the metabolism of drugs and other xenobiotics. Biomedical Chromatography. 2000;**14**:351-372

[66] Olah TV. The development and implementation of bioanalytical methods using LC-MS to support ADME studies in early drug discovery and candidate selection. In: Ernst Schering Research Foundation Workshop. Berlin, Heidelberg: Springer; 2002. pp. 155-183

[67] Lim CK, Lord G. Current developments in LC-MS for

pharmaceutical analysis. Biological & Pharmaceutical Bulletin. 2002;**25**(5):547-557

[68] Jenkins KM, Angeles R, Quintos MT, Xu R, Kassel DB, Rourick RA. Automated high throughput ADME assays for metabolic stability and cytochrome P450 inhibition profiling of combinatorial libraries. Journal of Pharmaceutical and Biomedical Analysis. 2004;**34**(5):989-1004

[69] Thompson TN. Early ADME in support of drug discovery: The role of metabolic stability studies. Current Drug Metabolism. 2000;**1**(3):215-241

[70] Hopfgartner G, Husser C, Zell M. High-throughput quantification of drugs and their metabolites in biosamples by LC-MS/MS and CEMS/MS: Possibilities and limitations. Therapeutic Drug Monitoring. 2002;**24**:134-143

[71] Cox K. Special requirements for metabolite characterization. In: Using Mass Spectrometry for Drug Metabolism Studies. Boca Raton: CRC Press; 2005. pp. 229-252

[72] Clarke NJ, Rindgen D, Korfmacher WA, Cox KA. Systematic LC/MS metabolite identification in drug discovery. Analytical Chemistry. 2001;**73**(15):430A-439A

[73] Hopfgartner G, Husser C, Zell M. Rapid screening and characterization of drug metabolites using a new quadrupole–linear ion trap mass spectrometer. Journal of Mass Spectrometry. 2003;**38**:138-150

[74] Leclercq L, Delatour C, Hoes I, Brunelle F, Labrique X. Use of a five-channel multiplexed electrospray quadrupole time-of-flight hybrid mass spectrometer for metabolite identification. Rapid Communications in Mass Spectrometry. 2005;**19**:1611-1618

[75] Bateman KP, Kellmann M, Muenster H, Papp R, Taylor L. Quantitative-qualitative data acquisition using a benchtop Orbitrap mass spectrometer. Journal of the American Society for Mass Spectrometry. 2009;**8**:1441-1450

[76] Chernushevich IV, Loboda AV, Thomson BA. An introduction to quadrupole-time-of flight mass spectrometry. Journal of Mass Spectrometry. 2001;**36**:849-865

[77] Hu Q, Noll RJ, Li H, Makarov A, Hardman M, Cooks RG. The Orbitrap: A new mass spectrometer. Journal of Mass Spectrometry. 2005;**40**:430-443

[78] Lim HK, Chen J, Sensenhauser C, Cook K, Subrahmanyam V. Metabolite identification by data-dependent accurate mass spectrometric analysis at resolving power of 60,000 in external calibration mode using an LTQ/Orbitrap. Rapid Communications in Mass Spectrometry. 2007;**21**:1821-1832

[79] Makarov A, Denisov E, Kholomeev A, Balschun W, Lange O, Strupat K, et al. Performance evaluation of a hybrid linear ion trap/orbitrap mass spectrometer. Analytical Chemistry. 2006;**78**(7):2113-2120

[80] Fernandez-Metzler CL, Owens KG, Baillie TA, King RC. Rapid liquid chromatography with tandem mass spectrometry-based screening procedures for studies on the biotransformation of drug candidates. Drug Metabolism and Disposition. 1999;**27**(1):32-40

[81] Mahajan MK, Evans CA. Dual negative precursor ion scan approach for rapid detection of glutathione conjugates using liquid chromatography/tandem mass spectrometry. Rapid Communications in Mass Spectrometry. 2008;**22**(7):1032-1040

[82] Dieckhaus CM, Fernandez-Metzler CL, King R, Krolikowski PH, Baillie TA. Negative ion tandem mass spectrometry for the detection of glutathione conjugates. Chemical Research in Toxicology. 2005;**18**(4):630-638

[83] Lin JH, Lu AY. Role of pharmacokinetics and metabolism in drug discovery and development. Pharmacological Reviews. 1997;**49**(4):403-449

[84] Alavijeh MS, Palmer AM. The pivotal role of drug metabolism and pharmacokinetics in the discovery and development of new medicines. IDrugs. 2004;7(8):755-763

[85] Ackermann BL, Berna MJ, Murphy AT. Recent advances in use of LC/MS/MS for quantitative high-throughput bioanalytical support of drug discovery. Current Topics in Medicinal Chemistry. 2002;**2**:53-66

[86] Hopfgartner G, Bourgogne E. Quantitative high-throughput analysis of drugs in biological matrices by mass spectrometry. Mass Spectrometry Reviews. 2003;**22**(3):195-214

[87] Cox KA, White RE, Korfmacher WA. Rapid determination of pharmacokinetic properties of new chemical entities: in vivo approaches. Combinatorial Chemistry & High Throughput Screening. 2002;**5**(1):29-37

[88] Ong VS, Cooke KL, Kosara CM, Brubaker WF. Quantitative bioanalysis: An integrated approach for drug discovery and development. International Journal of Mass Spectrometry. 2004;**238**:139-152

[89] Manitpisitkul P, White RE. Whatever happened to cassette-dosing pharmacokinetics? Drug Discovery Today. 2004;**9**(15):652-658

[90] Ohkawa T, Ishida Y, Kanaoka E, Takahashi K, Okabe H. A new generic column switching system for quantitation in cassette dosing using LC/MS/MS. Journal of Pharmaceutical and Biomedical Analysis. 2003;**31**:1089-1099

[91] Zhang MY, Kerns E, McConnell O, Sonnenberg-Reines J, Zaleska MM. Brain and plasma exposure profiling in early drug discovery using cassette administration and fast liquid chromatography-tandem mass spectrometry. Journal of Pharmaceutical and Biomedical Analysis. 2004;**34**:359-368

[92] Sadagopan N, Pabst B, Cohen L. Evaluation of online extraction/mass spectrometry for in vivo cassette analysis. Journal of Chromatography. B, Analytical Technologies in the Biomedical and Life Sciences. 2005;**820**(1):59-67

[93] White RE, Manitpisitkul P. Pharmacokinetic theory of cassette dosing in drug discovery screening. Drug Metabolism and Disposition. 2001;**29**(7):957-966

[94] Hsieh Y, Wang G, Wang Y, Chackalamannil S, Brisson JM. Simultaneous determination of a drug candidate and its metabolite in rat plasma samples using ultrafast monolithic column high-performance liquid chromatography/tandem mass spectrometry. Rapid Communications in Mass Spectrometry. 2002;**16**:944-950

[95] Jemal M, Ouyang Z. Enhanced resolution triple-quadrupole mass spectrometry for fast quantitative bioanalysis using liquid chromatography/tandem mass spectrometry: Investigations of parameters that affect ruggedness. Rapid Communications in Mass Spectrometry. 2003;**17**:24-38

[96] Romanyshyn L, Tiller PR, Alvaro R, Pereira A, Hop CE. Ultra-fast gradient vs fast isocratic chromatography in bioanalytical quantification by liquid chromatography/tandem mass spectrometry. Rapid Communications in Mass Spectrometry. 2001;**15**:31-319

[97] Hsieh Y, Chintala M, Mei H, Agans J, Brisson JM. Quantitative screening and matrix effect studies of drug discovery compounds in monkey plasma using fast-gradient liquid chromatography/tandem mass spectrometry. Rapid Communications in Mass Spectrometry. 2001;**15**:2481-2487

[98] Dunn-Meynell KW, Wainhaus S, Korfmacher WA. Optimizing an ultrafast generic high-performance liquid chromatography/tandem mass spectrometry method for faster discovery pharmacokinetic sample throughput. Rapid Communications in Mass Spectrometry. 2005;**19**:2905-2910

[99] Tanak N, Kobayashi H, Ishizuka N, Minakuchi H, Nakanishi K, Hosoya K, et al. Monolithic silica columns for high-efficiency chromatographic separations. Journal of Chromatography. A. 2002;**965**(1-2):35-49

[100] Tanaka N, Kobayashi H, Nakanishi K, Minakuchi H, Ishizuka N. Monolithic LC columns. Analytical Chemistry. 2001;**73**(15):420A-429A

[101] Rozenbrand J, van Bennekom WP, Unger KK, de Jong GJ. Fast LC separation of a myoglobin digest: A case study using monolithic and particulate RP 18 silica capillary columns. Analytical and Bioanalytical Chemistry. 2006;**385**(6):1055-1061

[102] Korfmacher WA, Veals J, Dunn-Meynell K, Zhang X, Tucker G. Demonstration of the capabilities of a parallel high performance liquid chromatography tandem mass spectrometry system for use in the analysis of drug discovery plasma samples. Rapid Communications in Mass Spectrometry. 1999;**13**:1991-1998

[103] Jemal M, Huang M, Mao Y, Whigan D, Powell ML. Increased throughput in quantitative bioanalysis using parallel-column liquid chromatography with mass spectrometric detection. Rapid Communications in Mass Spectrometry. 2001;**15**:994-999

[104] Wu JT. The development of a staggered parallel separation liquid chromatography/tandem mass spectrometry system with on-line extraction for high-throughout screening of drug candidates in biological fluids. Rapid Communications in Mass Spectrometry. 2001;**15**:73-81

[105] Taylor PJ. Matrix effects: The Achilles heel of quantitative high-performance liquid chromatography–electrospray–tandem mass spectrometry. Clinical Biochemistry. 2005;**38**(4):328-334

[106] Hall TG, Smukste I, Bresciano KR, Wang Y, McKearn D, Savage RE. Identifying and overcoming matrix effects in drug discovery and development. In: Tandem Mass Spectrometry-kApplications and Principles. Rijeka: InTech; 2012

[107] Bakhtiar R, Majumdar TK. Tracking problems and possible solutions in the quantitative determination of small molecule drugs and metabolites in biological fluids using liquid chromatography–mass spectrometry. Journal of Pharmacological and Toxicological Methods. 2007;**55**(3):227-243

[108] Côté C, Bergeron A, Mess J-N, Furtado M, Garofolo F. Matrix effect elimination during LC–MS/MS bioanalytical method development. Bioanalysis. 2009;**1**(7):1243-1257

[109] Trufelli H, Palma P, Famiglini G, Cappiello A. An overview of matrix effects in liquid chromatography–mass spectrometry. Mass Spectrometry Reviews. 2011;**30**(3):491-509

[110] Vijaya BV. Identification and reduction of matrix effects caused by polyethylene glycol 400 in bioanalysis using liquid chromatography/tandem mass spectrometry. International Journal of Pharmaceutical Innovations. 2013;**3**(1):48-65

[111] Vijaya Bhaskar V, Middha A, Tiwari S, Shivakumar S. Identification and reduction of matrix effects caused by cremophor EL in bioanalysis using liquid chromatography/tandem mass spectrometry. Journal of Analytical and Bioanalytical Tehcniques. 2013;**4**(3):1-7

[112] Vijaya Bhaskar V, Sudhir T, Anil M, Savithiri S. Identification and reduction of matrix effects caused by Solutol Hs15 in bioanalysis using liquid chromatography/tandem mass spectrometry. Journal of Analytical and Bioanalytical Techniques. 2013;**4**(166):1-8

[113] Tong XS, Wang J, Zheng S, Pivnichny JV, Griffin PR, Shen X, et al. Effect of signal interference from dosing excipients on pharmacokinetic screening of drug candidates by liquid chromatography/mass spectrometry. Analytical Chemistry. 2002;**74**(24):6305-6313

[114] Shou WZ, Naidong W. Post-column infusion study of the 'dosing vehicle effect'in the liquid chromatography/tandem mass spectrometric analysis of discovery pharmacokinetic samples. Rapid Communications in Mass Spectrometry. 2003;**17**(6):589-597

[115] Tiller PR, Romanyshyn LA. Implications of matrix effects in ultra-fast gradient or fast isocratic liquid chromatography with mass spectrometry in drug discovery. Rapid Communications in Mass Spectrometry. 2002;**16**(2):92-98

[116] Vijaya Bhaskar V, Anil M. Liquid chromatography/tandem mass spectrometry method for quantitation of Cremophor EL and its applications. International Journal of Analytical Chemistry. 2013;**2013**:1-11

[117] Vijaya Bhaskar V, Middha A, Tiwari S, Shivakumar S. Liquid chromatography/tandem mass spectrometry method for quantitative estimation of polyethylene glycol 400 and its applications. Journal of Chromatography B. 2013;**926**:68-76

[118] Vijaya Bhaskar V, Middha A, Tiwari S, Shivkumar S. Determination of Cremophor EL in rat plasma by LC-MS/MS: Application to a pharmacokinetic study. Journal of Analytical and Bioanalytical Tehcniques. 2013;**4**:163

[119] Vijaya Bhaskar V, Middha A, Srivastava P, Rajagopal S. Liquid chromatography/tandem mass spectrometry method for quantitative estimation of solutol HS15 and its applications. Journal of Pharmaceutical Analysis. 2015;**5**(2):120-129

[120] Cech NB, Enke CG. Practical implications of some recent studies in electrospray ionization fundamentals. Mass Spectrometry Reviews. 2001;**20**(6):362-387

[121] King R, Bonfiglio R, Fernandez-Metzler C, Miller-Stein C, Olah T. Mechanistic investigation of ionization suppression in electrospray ionization. Journal of the American Society for Mass Spectrometry. 2000;**11**(11):942-950

[122] Bennett P, Liang H, editors. Overcoming Matrix Effects Resulting from Biological Phospholipids through Selective Extractions in Quantitative LC/MS/MS. 52nd ASMS Conference

on Mass Spectrometry. Nashville, Tennesee; 2004

[123] Chambers E, Wagrowski-Diehl DM, Lu Z, Mazzeo JR. Systematic and comprehensive strategy for reducing matrix effects in LC/MS/MS analyses. Journal of Chromatography B. 2007;**852**(1-2):22-34

[124] Bonfiglio R, King RC, Olah TV, Merkle K. The effects of sample preparation methods on the variability of the electrospray ionization response for model drug compounds. Rapid Communications in Mass Spectrometry. 1999;**13**(12):1175-1185

[125] Van Hout M, Niederländer H, De Zeeuw R, De Jong G. Ion suppression in the determination of clenbuterol in urine by solid-phase extraction atmospheric pressure chemical ionisation ion-trap mass spectrometry. Rapid Communications in Mass Spectrometry. 2003;**17**(3):245-250

[126] Schuhmacher J, Zimmer D, Tesche F, Pickard V. Matrix effects during analysis of plasma samples by electrospray and atmospheric pressure chemical ionization mass spectrometry: Practical approaches to their elimination. Rapid Communications in Mass Spectrometry. 2003;**17**(17):1950-1957

[127] Larger PJ, Breda M, Fraier D, Hughes H, James CA. Ion-suppression effects in liquid chromatography–tandem mass spectrometry due to a formulation agent, a case study in drug discovery bioanalysis. Journal of Pharmaceutical and Biomedical Analysis. 2005;**39**(1-2):206-216

[128] Renew JE, Huang C-H. Simultaneous determination of fluoroquinolone, sulfonamide, and trimethoprim antibiotics in wastewater using tandem solid phase extraction and liquid chromatography–electrospray mass spectrometry. Journal of Chromatography. A. 2004;**1042**(1-2): 113-121

Section 2

Applications of Mass Spectrometer

Chapter 4

Application of Two-Dimensional Gel Electrophoresis in Combination with Mass Spectrometry in the Study of Hormone Proteoforms

Xianquan Zhan and Tian Zhou

Abstract

Hormone is a category of important endocrine regulatory proteins in human endocrine systems. Clarification of hormone proteoforms directly leads to understanding of its biological roles. Two-dimensional gel electrophoresis (2DGE) in combination with mass spectrometry (MS) plays important roles in identification of hormone proteoforms such as human growth hormone (hGH) proteoforms and human prolactin (hPRL) proteoforms. This book chapter will review the hormone proteoforms focusing on hGH and hPRL, the methodology of hormone proteoform study, and future perspective of human hormone proteoform study to find biomarkers for in-depth understanding of molecular mechanisms, and individualized and precise diagnosis, therapy, and prognostic assessment of hormone-related diseases.

Keywords: two-dimensional gel electrophoresis, mass spectrometry, tandem mass spectrometry, liquid chromatography, hormone, growth hormone, prolactin, variant, proteoform

1. Introduction

Hormone is a category of important endocrine regulatory proteins in the human endocrine systems. Hormone is a chemical message substance synthesized by highly differentiated endocrine cells and directly secreted into the blood, which has high biological activities and transmits information in the body as a messenger. It is a vital substance in human life and plays important roles in regulating physiological processes such as metabolism, growth, and development.

Notably, human growth hormone (hGH) and human prolactin (hPRL) are two key hormones in human body. The hGH, also known as somatotropin, is produced in the acidophilic somatotroph cells of the anterior pituitary gland and is a 191 amino acid single chain polypeptide, which is released into the blood circulation and takes part in the hypothalamic-anterior pituitary-skeletal muscle axis system to regulate growth and development in human body [1, 2]. The synthesis and release of hGHs are affected by multiple complex feedback mechanisms, and the major regulators are growth hormone-releasing hormone (GHRH) synthesized in the hypothalamus, somatostatin derived from various tissues of the body, and ghrelin produced in the gastrointestinal tract. The mechanism of the effect of hGH is to directly bind with target organs for response of stimulation or it is indirectly influenced by the role of insulin-like growth factor-1 (IGF-1). The IGF-1 secreted from hepatocytes responds to the binding of hGH to surface receptors. If the Janus activating tyrosine kinases (JAKs) 1 and 2 are activated and bound to the cytoplasmic transcriptions factors STAT1, STAT3, and STAT5, and transported into the nucleus, thus enhanced gene transcription and metabolism would be induced, and the corresponding IGF-1 is generated and released into circulation. IGF-1 affects the growth and metabolism of peripheral tissues, which is able to understand as a combined impact of IGF-1 and hGH [2, 3]. The deficiency of hGH leads to growth deficit in children and the GH deficiency syndrome in adults, conversely, hGH hypersecretion results in gigantism or acromegaly [4–7]. It is obvious that hGH plays an important role in regulation of human life activities. In addition, hPRL should be paid more attention to its relevant functional study. The hPRL, a four long α-helix protein hormone, is mainly secreted by lactotrophs in the anterior pituitary gland except placental PRL [8]. The secretion of hPRL is usually regulated by PRL inhibitors such as dopamine from hypothalamus [9]. The hPRL is a polypeptide hormone with a wide variety of functions such as controls osmotic pressure and vascularization, participates in the immune response, and promotes neurogenesis in maternal and fetal brains [10, 11]. Moreover, the hPRL transports to the target organs and tissues via the blood circulation to bind to two different types of long or short hPRL receptors (hPRLRs) to activate signal pathways including JaK2 activation, Ras-Raf-MAPK pathway, modulatory pathways, PI3K and downstream pathways, and Stats [12]. However, hGH and hPRL exist multiple structural and functional formats, namely different proteoforms. The proteoform can clarify the function and effect of specific protein in human body and is an intuitive and visual expression of the gene. Therefore, the clarification of hormone proteoforms is necessary to understand its biological roles. Protein proteoforms are primarily derived from alternative splicing, post-translational modifications (PTMs), translocation, redistribution, and spatial conformation alterations [13]. The protein proteoforms can be identified with two-dimensional gel electrophoresis (2DGE) and mass spectrometry (MS).

The isoelectric point (*pI*) and relative mass (*Mr*) are the basic characteristics of a proteoform. 2DGE is depended on different *pI* and *Mr* values to separate proteoforms, which first separates proteoforms by different *pI* values in the isoelectric focusing (IEF) direction, and second separates proteoforms by different *Mr* values in the sodium dodecyl sulfate-polyacrylamide gel electrophoresis (SDS-PAGE) direction [14]. 2DGE-based Western blotting coupled with specific antibody can detect a proteoform of a specific protein. Thus, the proteoforms of hGH and hPRL are able to be separated and arrayed with 2DGE. MS is an effect method to characterize protein proteoforms and identify PTMs with an analysis of amino acid sequence and determination of PTM sites [15]. The combination of 2DGE and MS plays an important role in detection, identification, and quantification of proteoforms of hGH and hPRL. The MS methods that were used to study the proteoforms of hGH and hPRL were commonly divided into peptide fingerprint (PMF) and tandem mass spectrometry (MS/MS) analyses. PMF data are commonly obtained with matrix-assisted laser desorption/ionization (MALDI)-time-of-flight (TOF) mass spectrometey (MS), and MS/MS data can be obtained from MALDI-TOF-TOF MS or liquid chromatography (LC)-electrospray ionization (ESI)-quadruple-ion trap (Q-IT) MS.

This book chapter will mainly review the hormone proteoforms focusing on hGH and hPRL, the methodology of hormone proteoform study, and future perspective to find effective and potential hormone proteoform biomarkers for in-depth understanding of molecular mechanisms, and individualized and precise diagnosis, therapy, and prognostic assessment of hormone-related diseases.

2. Materials and methods

2.1 Tissues and protein extraction

The human control pituitary tissue samples were post-mortem tissues, and human pituitary adenoma tissue samples were obtained from neurosurgery. The detailed information of those tissue samples were described previously [15, 16]. The collected tissues were immediately frozen in liquid nitrogen, and stored at −80°C until used. The protein extraction of pituitary control and adenoma tissues was performed as described previously [17, 18]. In brief, the contaminated blood in each tissue sample was washed with 0.9% NaCl (3 mL, 5×). The proteins were extracted with protein extraction buffer that consists of 2 mol/L thiourea, 7 mol/L urea, 40 g/L CHAPS, 100 mmol/L dithiothreitol (DTT), 5 mol/L immobilized pH gradient (IPG) buffer pH 3–10 NL, and a trace of bromophenol blue, followed by centrifugation (15,000×*g*, 15 min, 4°C). The supernatant was collected as protein sample, and its protein concentration was determined with a Bio-Rad 2D Quant kit (Bio-Rad) [15].

2.2 2DGE and 2DGE-based Western blot

Each protein sample was first separated by IEF with pH 3–10 NL IPG strips (180 mm × 3 mm × 0.5 mm), under the IEF condition that was a gradient from 0 to 250 V within 1 h (125 Vh), a gradient from 125 to 1000 V within 1 h (500 Vh), a gradient from 1000 to 8000 V within 1 h (4000 Vh), a step-and-hold at 8000 V for 4 h (32,000 Vh), and a step-and-hold at 500 V for 0.5 h (250 Vh) to achieve a total of 36,875 Vh within ~7.5 h [19]. After IEF, proteins were reduced with DTT, and alkylated with iodoacetamide, and then were separated with the 12% SDS-PAGE resolving gel (250 mm × 215 mm × 1.0 mm) in the Tris-glycine-SDS electrophoresis buffer that contained 192 mmol/L glycine, 25 mmol/L Tris-base, and 1 g/L SDS with an electrophoresis condition (250 V, 360 min, 25°C) [20]. The 2DGE-separated proteins were visualized with silver-staining [21].

After 2DGE, the proteins were transferred onto a polyvinylidene fluoride (PVDF) membrane, blocked with bovine serum albumin (BSA), and incubated with hormone antibodies and secondary antibody. The proteins on the membrane were visualized with 5-bromo-4-chloro-3-indolyl phosphate [15].

The silver-stained 2D gels and 2DE-Western blot images were digitized and analyzed with Discovery Series PDQuest 2D Gel Analysis software [22, 23].

2.3 MS-characterization

2.3.1 MALDI-TOF MS

The tryptic peptide mixture was purified with ZipTipC18, and analyzed with a Perspective Biosystems MALDI-TOF Voyager DE-RP MS (Framingham, MA, USA). The PMF data were obtained to search Swiss-Prot database with PeptIdent software for protein identification. The detailed procedure was described previously [22].

2.3.2 LC-ESI-Q-IT MS

The tryptic peptide mixture was purified with ZipTipC18, and analyzed with LC-ESI-Q-IT MS on an LCQ^{Deca} mass spectrometer (Thermo Finnigan, San Jose, CA, USA). The MS/MS data were obtained to search Swiss-Prot database for protein identification. The detailed procedure was described previously [22].

2.3.3 MALDI-TOF-TOF MS

The tryptic peptide mixture was purified with ZipTipC18, and analyzed with MALDI-TOF-TOF MS on Perspective Biosystems UltraFlex III MALDI-TOF-TOF (Bruker Daltonics). The MS and MS/MS data were obtained to search Swiss-Prot database for protein identification. The detailed procedure was described previously [15].

2.4 Bioinformatics

Many phosphorylation sites of hGH proteoforms were identified with MS/MS, and deamidation was found in many hGH proteoforms. The PTM sites of hPRL in human pituitary adenoma and control tissues, including phosphorylation sites, N-glycosylation sites, and O-glycosylation sites were predicted with NetPhos 3.1 Server (http://www.cbs.dtu.dk/services/NetPhos) [24, 25], NetNGlyc 1.0 Server (http://www.cbs.dtu.dk/services/NetNGlyc) [26], and NetOGlyc 4.0 Server (http://www.cbs.dtu.dk/services/NetOGlyc) [27], respectively.

3. Results and discussion

3.1 MS-identification of open reading frames (ORFs) of hGH and hPRL proteoforms within 2DGE pattern, and confirmed with Western blot

Over 1000 protein spots were found in each pituitary 2D gel (n = 30). MS is an effective method to identify ORFs of a gene across 2DGE map, generally cannot identify proteoforms with the common procedure because MS only detects a partial amino acid sequence but not its complete sequence of a protein. However, each proteoform has its specific *pI* and *Mr*, which can be separated with 2DGE. 2DGE coupled with MS can effectively array and identify proteoforms that are derived from the same gene. Thus, 24 hGH proteoforms (**Figure 1A**) [1] and 6 hPRL proteoforms (**Figure 1B**) [15] were MS-characterized within pituitary 2DGE map.

Some hGH proteoforms and hPRL proteoforms within 2DGE map (**Figure 1**) were also validated with Western immunoblot. For example, four hPRL proteoforms were confirmed with 2DGE-Western blot (**Figure 2**). It demonstrated that 2DGE-MS was an effective and reliable approach to detect and identify proteoforms derived from hGH and hPRL genes. The other hGH proteoforms and hPRL proteoforms were not be validated with immnuoblot, which might be due to no-reactivity of antibody to a specific proteoform.

The hGH proteoforms or hPRL proteoforms in each 2D spot were identified with MS including PMF and MS/MS. For example, the protein in spot 6 in **Figure 1A** was identified as hGH isoform 1 (P01241) with MALDI-TOF PMF data (**Figure 3**). The protein in spot v6 in **Figure 1B** was identified as hPRL (P01236) with MALDI-TOF-TOF PMF data (**Figure 4**), and MS/MS data (**Figure 5**).

Therefore, 2DGE-MS clearly identified 24 hGH proteoforms and 6 hPRL proteoforms with different pI-M_r distributions on a 2DGE pattern. Furthermore,

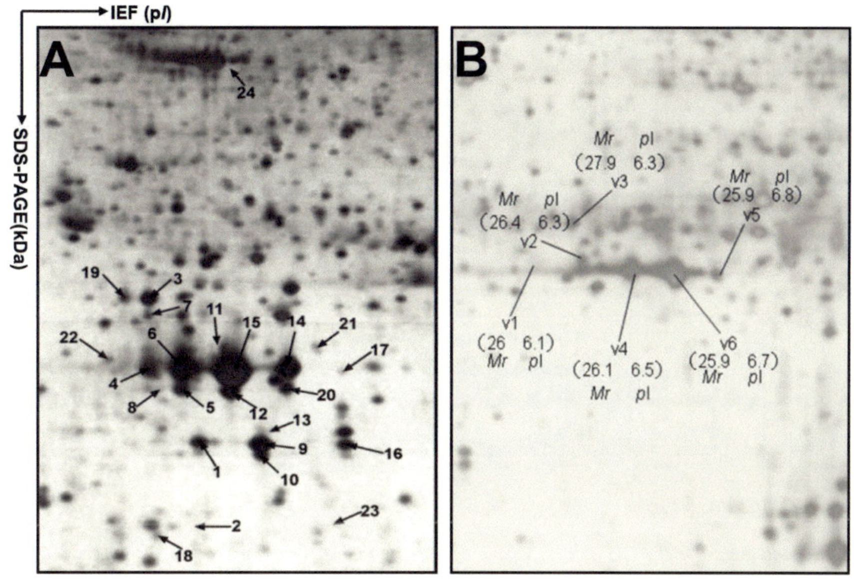

Figure 1.
2DGE patterns of hGH proteoforms (A) and hPRL proteoforms (B) in human pituitaries. The total proteins extracted from pituitary tissues were separated with IPG strip pH 3–10 NL and 12% SDS-PAGE gel, and stained with silver-staining. Modified from Zhan et al. [1], with permission from Wiley-VCH, copyright 2005; and from Qian et al. [15], with permission from Frontiers publisher open access article, copyright 2018.

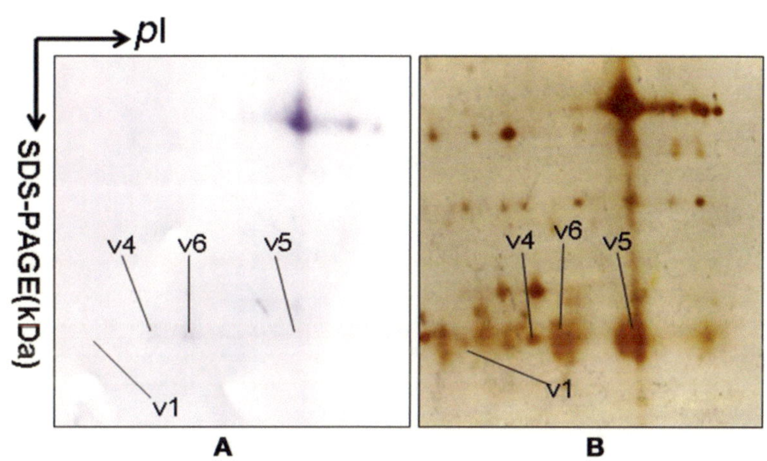

Figure 2.
hPRL proteoforms were validated with 2DGE-Western blot analysis. (A) Western blot image of hPRL proteoforms (rabbit anti-hPRL antibodies + goat anti-rabbit alkaline phosphatase-conjugated IgG). (B) Silver-stained image on a 2D gel after proteins were transferred to a PVDF membrane. Reproduced from Qian et al. [15], with permission from Frontiers publisher open access article, copyright 2018.

one investigated the reasons to form those hormone proteoforms, including signal peptide, splicing variation, and PTMs such as deamination, phosphorylation, and glycosylation.

3.2 MS-determination of signal peptide contained in each hGH or hPRL proteoform in human pituitaries

If signal peptide is contained in the sequence of hGH or hPRL, then it means that hormone is a prohormone (**Figure 6**). It is necessary to determine whether signal

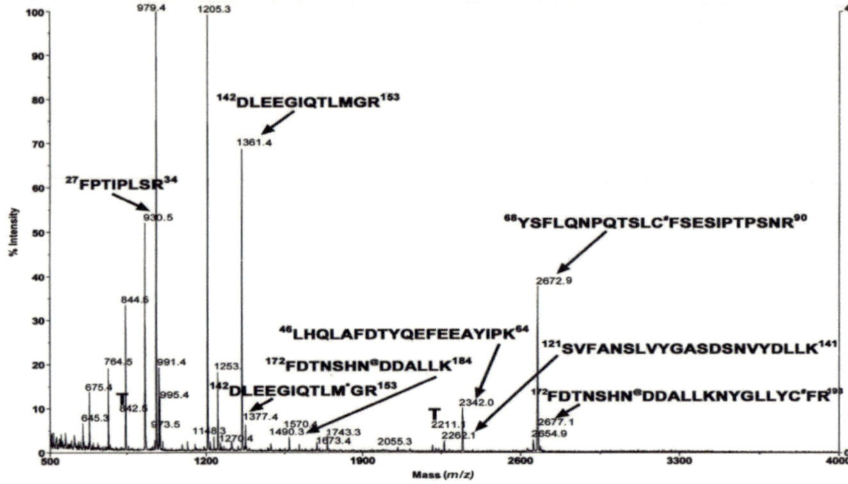

Figure 3.
MALDI-TOF MS spectrum of hGH isoform 1 (spot 6). T = autodigested fragments of trypsin. M = oxidized Met. C# = carbamidomethyl-Cys. N@ = deamidated Asn. Reproduced from Zhan et al. [1], with permission from Wiley-VCH, copyright 2005.*

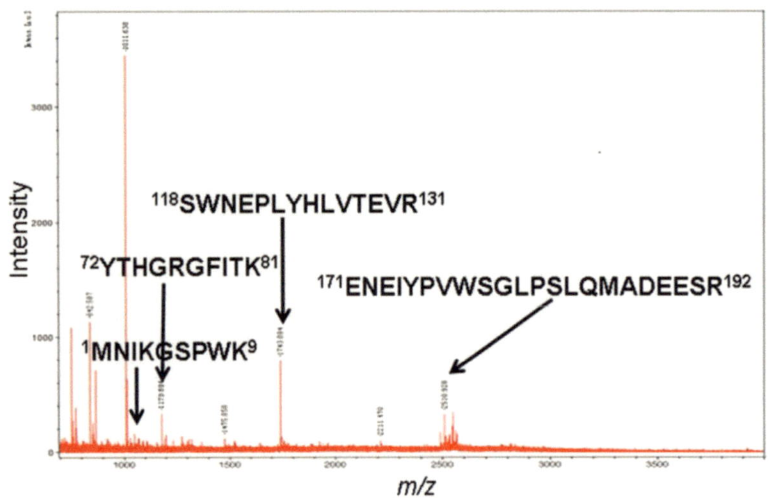

Figure 4.
MALDI-TOT-TOF MS spectrum of hPRL (spot v6). Reproduced from Qian et al. [15], with permission from Frontiers publisher open access article, copyright 2018.

peptide is contained in each hGH proteoform or hPRL proteoform. MS can identify the characteristic tryptic peptide sequences before and after removal of signal peptide in hGH or hPRL (**Table 1**). For 24 hGH proteoforms in human pituitaries, checking all MS data of those 24 hGH proteoforms, the tryptic peptide sequence FPTIPLSR (position 27–34) (**Figure 3**) was detected for each hGH proteoform, but no any characteristic tryptic peptide sequence before removal of signal peptide was found (**Table 1**). It clearly demonstrated that no signal peptide sequence was contained in those 24 hGH proteoforms in the 2DGE pattern (**Figure 1A**). For 6 hPRL proteoforms in human pituitaries, checking all MS data of those 6 hPRL proteoforms, the tryptic peptide sequence MNIKGSPWK (position 1–9) (**Figure 4**) was detected for each hPRL proteoform, but no any characteristic tryptic peptide sequence after removal of signal peptide was found (**Table 1**). It clearly demonstrated that signal peptide sequence was contained in those 6 hPRL proteoforms in the 2DGE pattern (**Figure 1B**).

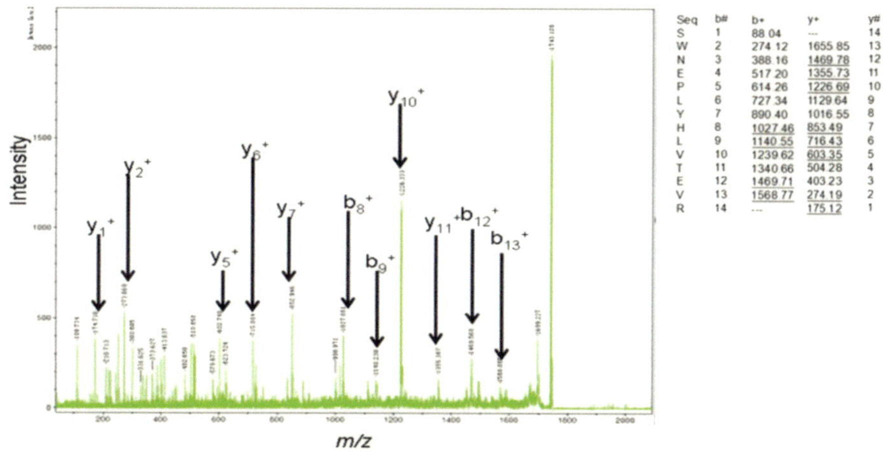

Figure 5.
MALDI-TOF-TOF MS/MS spectrum of the tryptic peptide [118]SWNEPLYHLVTEVR[131] from hPRL in spot v6. Reproduced from Qian et al. [15], with permission from Frontiers publisher open access article, copyright 2018.

	1	11	21	31	41	
hGH						
1	**MATGSRTSLL**	**LAFGLLCLPW**	**LQEGSA**FPTI	PLSRLFDNAM	LRAHRLHQLA	50
51	FDTYQEF**EEA**	**YIPKEQKYSF**	**LQ**NPQTSLCF	SESIPTPSNR	EETQQKSNLE	100
101	LLRISLLLIQ	**SWLEPVQFLR**	**SVFANSLVYG**	**ASDSNVYDLL**	**KDLEEGIQTL**	150
151	**MGRLEDGSPR**	**TG**QIFKQTYS	KFDTNSHNDD	ALLKNYGLLY	CFRKDMDKVE	200
201	TFLRIVQCRS	VEGSCGF				
hPRL	1	11	21	31	41	
1	**MNIKGSPWKG**	**SLLLLLVSNL**	**LLCQSVAP**LP	ICPGGAARCQ	VTLRDLFDRA	50
51	VVLSHYIHNL	SSEMFSEFDK	RYTHGRGFIT	KAINSCHTSS	LATPEDKEQA	100
101	QQMNQKDFLS	LIVSILRSWN	EPLYHLVTEV	RGMQEAPEAI	LSKAVEIEEQ	150
151	TKRLLEGMEL	IVSQVHPETK	ENEIYPVWSG	LPSLQMADEE	SRLSAYYNLL	200
201	HCLRRDSHKI	DNYLKLLKCR	IIHNNNC			

Figure 6.
The amino acid sequences of hGH and hPRL prohormones. Modified from Zhan et al. [1], with permission from Wiley-VCH, copyright 2005; and from Qian et al. [15], with permission from Frontiers publisher open access article, copyright 2018.

3.3 MS-identification of splicing variants of hGH in human pituitaries

MS is an effective method to identify splicing variants of hGH. hGH splicing variant 2 is derived from deletion of positions 58–72 from hGH, hGH splicing variant 3 is derived from deletion of positions 111-148 from hGH, and hGH splicing variant 4 is derived from deletion of positions 117-162 from hGH (**Figure 6**). The characteristic tryptic peptides for each hGH splicing variant can be theoretically obtained and observed by MS (**Table 2**). The MS data of 24 hGH proteoforms were checked one-by-one. The results showed that hGH splicing variant 2 existed in spot 9, hGH splicing variant 3 existed in spots 2 and 23, and hGH splicing variant 4 existed in spot 13. hGH variant 1 existed in the rest 20 spots.

However, hPRL does not have splicing variants. The *Mr* of hPRL proteoforms was 26.0 kDa for v1, 26.1 kDa for v4, 25.9 kDa for v5, and 25.9 kDa for v6 in 2DGE pattern (**Figure 1**), which was almost equal to the *Mr* (25.9 kDa) of hPRL prohormone. Moreover, the *Mr* of hPRL proteoforms was 26.4 kDa for v2, and 27.9 kDa for v3, which is slightly greater than *Mr* (25.9 kDa) of hPRL prohormone. These data further confirmed that those six hPRL proteoforms were not generated from mature hPRL (position 29–227) but

Hormone	Before and after removal of signal peptide	Position	Characteristic tryptic peptide sequence	Calc. $[M + H]^+$	Observed $[M + H]^+$
hGH (signal peptide sequence 1–26)	Before	1–6	MATGSR	622.3 (MSO: 1; 638.3)	–
		7–34	TSLLLAFGLLCLPWLQEGSAFPTIPLSR	3043.7 3100.7(Cys_CAM: 17)	– –
		1–34	MATGSRTSLLLAFGLLCLPWLQEGSAFPTIPLSR	3646.9 3704.0 (Cys_CAM: 17) 3662.9 (MSO: 1)	– – –
		7–42	TSLLLAFGLLCLPWLQEGSAFPTIPLSRLFDNAMLR	4004.2 4061.2 (Cys_CAM: 17) 4020.1 (MSO: 40)	– – –
	After	27–34	FPTIPLSR	930.5	+
		27–42	FPTIPLSRLFDNAMLR	1891.0 1907.0 (MSO: 40)	± –

Hormone	Before and after removal of signal peptide	Position	Characteristic tryptic peptide sequence	Calc. $[M+H]^+$	Observed $[M+H]^+$
hPRL (signal peptide sequence 1–28)	Before	1–4	MNIK	505.2803	–
		1–9	MNIKGSPWK	1060.5608	+
		1–38	MNIKGSPWKGSLLLLLVSNL LLCQSVAPLPICPGGAAR	3930.1893	–
		5–9	GSPWK	574.2983	–
		5–38	GSPWKGSLLLLLVSNLLLCQ SVAPLPICPGGAAR	3443.9268	–
		10–38	GSLLLLLVSNLLLCQSVAPL PICPGGAAR	2888.6463	–
		10–44	GSLLLLLVSNLLLCQSVAPL PICPGGAARCQVTLR	3589.0154	–
	After	29–38	LPICPGGAAR	954.5189	–
		29–44	LPICPGGAARCQVTLR	1654.8879	–
		29–49	LPICPGGAARCQVTLRDLFD R	2301.1954	–

Modified from Zhan et al. [1], with permission from Wiley-VCH, copyright 2005; and from Qian et al. [15], with permission from Frontiers publisher open access article, copyright 2018. Note: +: this peptide ion was observed in each MS spectrum. −: this peptide ion was not observed by MS. ±: tryptic peptide ion was detected in some but not all PMF data from 24 hGH proteoforms.

Table 1.
Characteristic tryptic peptides before or after removal of signal peptide sequence in hGH or in hPRL.

GH variants	Tryptic peptide that covers the splicing site	Calculated [M + H] + ion (*m/z*)	Observed [M + H]$^+$ ion (*m/z*) in PMF data	hGH proteoforms (Spot No.)
1	^{142}DLEEGIQTLMGR^{153}	1361.7 1377.7 (MSO: 151)	One or two ions were detected to exclude splicing variants 3 and 4 One, two or three ions were detected to exclude splicing variant 2	1; 3–8; 10–12; 14–22; 24
	^{121}SVFANSLVYGASDSNVYDLLK^{141}	2262.1		
	^{46}LHQLA FDTYQEFEEAYIPK^{64}	2342.1		
	^{68}YSF LQNPQTSLCFSESIPTPSNR^{90}	2616.2 2673.2 (Cys-CAM: 79)		
	^{43}AHRLHQLAFDTYQ EFEEAYIPK^{64}	2706.3		
2	LHQLAFDTYQEF^{58}NPQTSLCFSESIPTPSNR	3470.6 3527.7 (Cys-CAM: 64)	— 3527.4	9
	AHRLHQLAFDTYQEF^{58}NPQTSLCFSESIPTPSNR	3834.8 3891.9 (Cys-CAM: 64)	— —	
	LHQLAFDTYQEF^{58}NPQTSLCFSESIPTPSNREETQQK	4213.9 4271.0 (Cys-CAM: 64)	— —	
3	ISLLLIQ^{111}TLMGR	1357.8 1373.8 (MSO: 113)	1357.7 —	2; 23
	ISLLLIQ^{111}TLMGRLEDGSPR	2112.2 2128.2 (MSO: 113)	— —	
	SNLE LLRISLLLIQ^{111}TLMGR	2183.3 2199.3 (MSO: 113)	— —	
4	ISLLLIQSWLEPV^{117}QIFK	2027.2	2027.0	13
	ISLLLIQSWLEPV^{117}QIFKQTYS K	2634.5	—	
	SNLE LLRISLLLIQSWLEPV^{117}QIFK	2852.7	—	

Modified from Zhan et al. [1], with permission from Wiley-VCH, copyright 2005.

Table 2.
Determination of the hGH splicing variants.

hGH proteoforms (Spot No.)	Splicing variants	Phosphorylation (p)	Deamidation (d)
1	1/2	pSer176	
2	3; 1/2		
3	1	pSer132	
4	1		
5	1	pSer176;	dAsn178
6	1	pSer77, pSer132, pSer176	dAsn178
7	1		
8	1	pSer176, pSer132	
9	2	pSer176	
10	1	pSer176	
11	1	pSer77, pSer132	
12	1	pSer176; pSer132	
13	4; 3		
14	1	pSer132	
15	1	pSer176, pSer132	
16	1	pSer176	dAsn178
17	1		
18	1	pSer176, pSer132	
19	1		
20	1		
21	1		
22	1		
23	3; 1		
24	1	pSer132	

Modified from Zhan et al. [1], with permission from Wiley-VCH, copyright 2005.

Table 3.
PTMs were identified in hGH proteoforms.

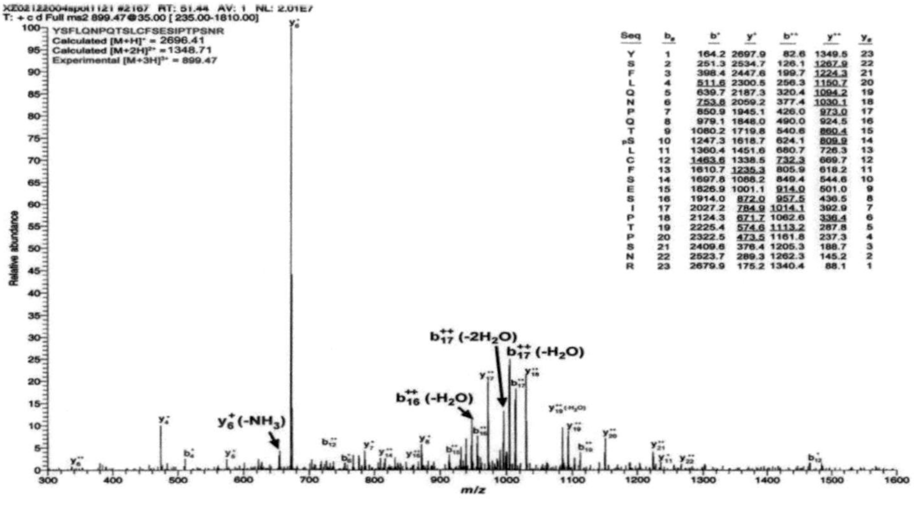

Figure 7.
MS² spectrum of a phosphopeptide ([M + 3H]³⁺: m/z = 899.47; retention time = 51.44 min; scan number 2167) derived from hGH proteoform 6 (Spot 6). Reproduced from Zhan et al. [1], with permission from Wiley-VCH, copyright 2005.

Sequence	#	x	Context	Score*	Kinase	Answer
Sequence	6	S	NIKGSPWKG	0.779	unsp	YES
Sequence	11	S	PWKGSLLLL	0.848	PKA	YES
Sequence	18	S	LLLVSNLLL	0.523	cdc2	YES
Sequence	42	T	RCQVTLRDL	0.891	unsp	YES
Sequence	61	S	IHNLSSEMF	0.718	unsp	YES
Sequence	62	S	HNLSSEMFS	0.553	unsp	YES
Sequence	66	S	SEMFSEFDK	0.991	unsp	YES
Sequence	72	Y	FDKRYTHGR	0.503	INSR	YES
Sequence	73	T	DKRYTHGRG	0.557	unsp	YES
Sequence	90	S	CHTSSLATP	0.585	DNAPK	YES
Sequence	93	T	SSLATPEDK	0.737	unsp	YES
Sequence	110	S	KDFLSLIVS	0.507	PKA	YES
Sequence	118	S	SILRSWNEP	0.749	unsp	YES
Sequence	124	Y	NEPLYHLVT	0.956	unsp	YES
Sequence	142	S	EAILSKAVE	0.517	CKII	YES
Sequence	151	T	IEEQTKRLL	0.983	unsp	YES
Sequence	163	S	ELIVSQVHP	0.623	ATM	YES
Sequence	169	T	VHTEPKENE	0.541	CKII	YES
Sequence	175	Y	ENEIYPVWS	0.804	unsp	YES
Sequence	191	S	ADEESRLSA	0.576	cdc2	YES
Sequence	194	S	ESRLSAYYN	0.982	unsp	YES
Sequence	207	S	LRRDSHKID	0.993	unsp	YES

Modified from Qian et al. [15], with permission from Frontiers publisher open access article, copyright 2018.
**Score > 0.5 means a statistically significant result.*

Table 4.
Prediction of phosphorylation sites in hPRL prohormone (positions 1–227) with NetPhos server with a score more than 0.5.

Seq name	Position	Potential	Jury agreement	N-Glyc result	
Sequence	2 NIKG	0.7530	**(9/9)**	+++	SEQUON ASN-XAA-SER/THR.
Sequence	19 NLLL	0.7151	**(9/9)**	++	
Sequence	59 NLSS	0.7380	**(9/9)**	++	
Sequence	84 NSCH	0.7312	(8/9)	+	
Sequence	104 NQKD	0.6020	(7/9)	+	
Sequence	120 NEPL	0.6051	(6/9)	+	
Sequence	172 NEIY	0.5346	(5/9)	+	
Sequence	198 NLLH	0.5642	(5/9)	+	
Sequence	212 NYLK	0.6726	(8/9)	+	
Sequence	224 NNNC	0.5146	(5/9)	+	
Sequence	225 NNC-	0.3576	(8/9)	—	
Sequence	226 NC--	0.3351	**(9/9)**	—	

Modified from Qian et al. [15], with permission from Frontiers publisher open access article, copyright 2018.
Note: Asparagines predicted to be N-glycosylated are highlighted in bold font.

Table 5.
Prediction of N-glycosylation sites in hPRL prohormone (positions 1–227) with NetNGlyc 1.0 server with score more than 0.5.

from hPRL prohormone (positions 1–227) in human pituitary tissues. These data coupled with MS data clearly demonstrated six hPRL proteoforms were not derived from splicing.

3.4 MS/MS-identification of PTMs in hormone proteoforms

MS/MS was an effective method to identify the PTMs and their modification sites of hGH and hPRL to clarify the reasons of formation of hGH proteoforms and hPRL proteoforms. MS/MS identified phosphorylation at sites Ser-77, Ser-132, and Ser-176 in many hGH proteoforms (**Table 3**). A representative MS/MS spectrum was shown to determine phosphorylation site Ser-77 in hGH proteoform 6 (Spot 6) (**Figure 7**). Also, deamidation was found in many hGH proteoforms. In addition, there would be other PTMs in hGH proteoforms that need to be further characterized.

#Seq name	Source	Feature	Start	End	Score	Strand	Frame	Comment
SEQUENCE	netOGlyc -4.0.0.13	CARBOHYD	25	25	0.134588	.	.	
SEQUENCE	netOGlyc -4.0.0.13	CARBOHYD	42	42	0.190888	.	.	
SEQUENCE	netOGlyc -4.0.0.13	CARBOHYD	54	54	0.194926	.	.	
SEQUENCE	netOGlyc -4.0.0.13	CARBOHYD	66	66	0.176466	.	.	
SEQUENCE	netOGlyc -4.0.0.13	CARBOHYD	73	73	0.111052	.	.	
SEQUENCE	netOGlyc -4.0.0.13	CARBOHYD	80	80	**0.613645**	.	.	#POSITIVE
SEQUENCE	netOGlyc -4.0.0.13	CARBOHYD	85	85	**0.618483**	.	.	#POSITIVE
SEQUENCE	netOGlyc -4.0.0.13	CARBOHYD	88	88	**0.602886**	.	.	#POSITIVE
SEQUENCE	netOGlyc -4.0.0.13	CARBOHYD	89	89	**0.717093**	.	.	#POSITIVE
SEQUENCE	netOGlyc -4.0.0.13	CARBOHYD	90	90	**0.928857**	.	.	#POSITIVE
SEQUENCE	netOGlyc -4.0.0.13	CARBOHYD	93	93	**0.778272**	.	.	#POSITIVE
SEQUENCE	netOGlyc -4.0.0.13	CARBOHYD	128	128	0.181904	.	.	
SEQUENCE	netOGlyc -4.0.0.13	CARBOHYD	151	151	0.11243	.	.	
SEQUENCE	netOGlyc -4.0.0.13	CARBOHYD	163	163	0.424529	.	.	
SEQUENCE	netOGlyc -4.0.0.13	CARBOHYD	169	169	0.122664	.	.	
SEQUENCE	netOGlyc -4.0.0.13	CARBOHYD	179	179	0.380532	.	.	
SEQUENCE	netOGlyc -4.0.0.13	CARBOHYD	183	183	0.309982	.	.	
SEQUENCE	netOGlyc -4.0.0.13	CARBOHYD	191	191	0.1589	.	.	
SEQUENCE	netOGlyc -4.0.0.13	CARBOHYD	194	194	0.249957	.	.	

Modified from Qian et al. [15], with permission from Frontiers publisher open access article, copyright 2018.
Note: The bold value means a statistically significantly positive result.

Table 6.
Prediction of O-glycosylation sites in hPRL prohormone (position 1–227) with NetOGlyc 4.0 server with score more than 0.5.

For hPRL proteoforms, bioinformatics including NetPhos 3.1 Server (http://www.cbs.dtu.dk/services/NetPhos) [24, 25] predicted 14 pS sites, 5 pT sites, and 3 pY sites in the hPRL (**Table 4**), NetNGlyc 1.0 Server (http://www.cbs.dtu.dk/services/NetNGlyc) [26] predicted ten significantly N-glycosylated sites (**Table 5**), and NetOGlyc 4.0 Server (http://www.cbs.dtu.dk/services/NetOGlyc) [27] predicted six significantly O-glycosylated sites in the hPRL (**Table 6**) in human pituitaries. These predicted PTM sites in hPRL proteoforms provided clues and needed to be confirmed with MS/MS in future studies.

4. Conclusions

hGH and hPRL are two important hormones in human endocrine systems, which are synthesized in the pituitary gland and secreted into the circulation system. Clarification of hGH proteoforms and hPRL proteoforms in human pituitary is essential step to elucidate their biological functions. Alternative splicing and PTMs are two important factors to cause proteoforms. 2DGE effectively presented 24 hGH proteoforms and 6 hPRL proteoforms with different pI-M_r distributions in 2DGE pattern of pituitary tissue proteome. MS/MS effectively identified their splicing variants and PTMs: (i) 24 hGH proteoforms in pituitary removed their signal peptide, whereas 6 hPRL proteoforms in human pituitary did not remove their signal peptide. (ii) 24 hGH proteoforms in human pituitary are derived from 4 types of alternative splicing variants, whereas 6 hPRL proteoforms do not exist any alternative splicing variants. (iii) PTMs pSer-77, pSer-132, and pSer-176 were identified in some of 24 hGH proteoforms, whereas although no PTMs were identified in hPRL proteoforms with MS/MS. However, phosphorylation, N-glycosylation, and O-glycosylation have been predicted with bioinformatics in hPRL proteoforms. Deamidation was presented in both hGH proteoforms and hPRL proteoforms. Therefore, 2DGE coupled with MS plays crucial roles in detection, identification, and quantification of hormone (hGH and hPRL) proteoforms, which benefits insight into the molecular mechanisms and discovery of effective biomarkers of hormone-related diseases.

Acknowledgements

The authors acknowledge the financial supports from the Xiangya Hospital Funds for Talent Introduction (to X.Z.), the Hunan Provincial "Hundred Talent Plan" program (to X.Z.), the Hunan Provincial Natural Science Foundation of China (Grant No. 14JJ7008 to X.Z.), China "863" Plan Project (Grant No. 2014AA020610-1 to X.Z.), and the National Natural Science Foundation of China (Grant Nos. 81572278 and 81272798 to X.Z.).

Conflict of interest

We declare that we have no financial and personal relationships with other people or organizations.

Author's contributions

X.Z. conceived the concept, designed the book chapter, wrote and critically revised the book chapter, coordinated and was responsible for the correspondence work and financial support. Z.T. participated in analysis of references and wrote partial manuscript.

Acronyms and abbreviations

BSA	bovine serum albumin
DTT	dithiothreitol
ESI	electrospray ionization
GHRH	growth hormone-releasing hormone
hGH	human growth hormone
hPRL	human prolactin
hPRLRs	hPRL receptors
IEF	isoelectric focusing
IPG	immobilized pH gradient
IGF-1	insulin-like growth factor-1
JAKs	janus-activating tyrosine kinases
LC	liquid chromatography
MALDI	matrix-assisted laser desorption/ionization
M_r	relative mass
MS	mass spectrometry
MS/MS	tandem mass spectrometry
ORFs	open reading frames
PAGE	polyacrylamide gel electrophoresis
pI	isoelectric point
PMF	peptide mass fingerprint
PTM	post-translational modification
PVDF	polyvinylidene fluoride
Q-IT	quadruple-ion trap
SDS-PAGE	sodium dodecyl sulfate-polyacrylamide gel electrophoresis
TOF	time-of-flight
2DGE	two-dimensional gel electrophoresis

Author details

Xianquan Zhan[1,2]* and Tian Zhou[1,2]

1 Key Laboratory of Cancer Proteomics of Chinese Ministry of Health, Xiangya Hospital, Central South University, Changsha, China

2 State Local Joint Engineering Laboratory for Anticancer Drugs, Xiangya Hospital, Central South University, Changsha, China

*Address all correspondence to: yjzhan2011@gmail.com

References

[1] Zhan X, Giorgianni F, Desiderio DM. Proteomics analysis of growth hormone isoforms in the human pituitary. Proteomics. 2005;5:1228-1241

[2] Brinkman JE, Sharma S. Physiology, Growth Hormone. StatPearls [Internet]. Treasure Island, FL: StatPearls Publishing; 2018

[3] Yoshizato H, Tanaka M, Nakai N, Nakao N, Nakashima K. Growth hormone (GH)-stimulated insulin-like growth factor I gene expression is mediated by a tyrosine phosphorylation pathway depending on C-terminal region of human GH receptor in human GH receptor-expressing Ba/F3 cells. Endocrinology. 2004;**145**:214-220

[4] Nwosu BU, Lee MM. Evaluation of short and tall stature in children. American Family Physician. 2008;**78**:597-604

[5] Melmed S. Mechanisms for pituitary tumorigenesis: The plastic pituitary. The Journal of Clinical Investigation. 2003;**112**:1603-1618

[6] Kato Y, Murakami Y, Sohmiya M, Nishiki M. Regulation of human growth hormone secretion and its disorders. Internal Medicine. 2002;**41**:7-13

[7] Kohler M, Thomas A, Püschel K, Schänzer W, Thevis M. Identification of human pituitary growth hormone variants by mass spectrometry. Journal of Proteome Research. 2009;**8**:1071-1076

[8] Pałubska S, Adamiak-Godlewska A, Winkler I, Romanek-Piva K, Rechberger T, Gogacz M. Hyperprolactinaemia–A problem in patients from the reproductive period to the menopause. Menopausal Review. 2017;**16**:1-7

[9] Tani N, Ikeda T, Watanabe M, Toyomura J, Ohyama A, Ishikawa T. Prolactin selectively transported to cerebrospinal fluid from blood under hypoxic/ischemic conditions. PLoS One. 2018;**13**:e0198673

[10] Shingo T, Gregg C, Enwere E, Fujikawa H, Hassam R, Geary C, et al. Pregnancy-stimulated neurogenesis in the adult female forebrain mediated by prolactin. Science. 2003;**299**:117-120

[11] Cabrera-Reyes EA, Limón-Morales O, Rivero-Segura NA, Camacho-Arroyo I, Cerbón M. Prolactin function and putative expression in the brain. Endocrine. 2017;**57**:199-213

[12] Clevenger CV, Furth PA, Hankinson SE, Schuler LA. The role of prolactin in mammary carcinoma. Endocrine Reviews. 2003;**24**:1-27

[13] Zhan X, Long Y, Lu M. Exploration of variations in proteome and metabolome for predictive diagnostics and personalized treatment algorithms: Innovative approach and examples for potential clinical application. Journal of Proteomics. 2018;**188**:30-40

[14] Zhan X, Yang H, Peng F, Li J, Mu Y, Long Y, et al. How many proteins can be identified in a 2DE gel spot within an analysis of a complex human cancer tissue proteome? Electrophoresis. 2018;**39**:965-980

[15] Qian S, Yang Y, Li N, Cheng T, Wang X, Liu J, et al. Prolactin variants in human pituitaries and pituitary adenomas identified with two-dimensional gel electrophoresis and mass spectrometry. Frontiers in Endocrinology. 2018;**9**:468

[16] Zhan X, Desiderio DM. Heterogeneity analysis of the human pituitary proteome. Clinical Chemistry. 2003;**49**:1740-1751

[17] Moreno CS, Evans CO, Zhan X, Okor M, Desiderio DM, Oyesiku NM.

Novel molecular signaling and classification of human clinically nonfunctional pituitary adenomas identified by gene expression profiling and proteomic analyses. Cancer Research. 2005;**65**:10214-10222

[18] Evans CO, Moreno CS, Zhan X, Mccabe MT, Vertino PM, Desiderio DM, et al. Molecular pathogenesis of human prolactinomas identified by gene expression profiling, RT-qPCR, and proteomic analyses. Pituitary. 2008;**11**:231-245

[19] Wang X, Guo T, Peng F, Long Y, Mu Y, Yang H, et al. Proteomic and functional profiles of a follicle-stimulating hormone positive human nonfunctional pituitary adenoma. Electrophoresis. 2015;**36**:1289-1304

[20] Guo T, Wang X, Li M, Yang H, Li L, Peng F, et al. Identification of glioblastoma phosphotyrosine-containing proteins with two-dimensional western blotting and tandem mass spectrometry. BioMed Research International. 2015;**2015**:134050

[21] Zhan X, Desiderio DM. Mass spectrometric identification of in vivo nitrotyrosine sites in the human pituitary tumor proteome. Methods in Molecular Biology. 2009;**566**:137-163

[22] Zhan X, Desiderio DM. A reference map of a human pituitary adenoma proteome. Proteomics. 2003;**3**:699-713

[23] Peng F, Li J, Guo T, Yang H, Li M, Sang S, et al. Nitroproteins in human astrocytomas discovered by gel electrophoresis and tandem mass spectrometry. Journal of the American Society for Mass Spectrometry. 2015;**26**:2062-2076

[24] Blom N, Gammeltoft S, Brunak S. Sequence and structure-based prediction of eukaryotic protein phosphorylation sites. Journal of Molecular Biology. 1999;**294**:1351-1362

[25] Blom N, Sicheritz-Ponten T, Gupta R, Gammeltoft S, Brunak S. Prediction of post-translational glycosylation and phosphorylation of proteins from the amino acid sequence. Proteomics. 2004;**4**:1633-1649

[26] Gupta R, Brunak S. Prediction of glycosylation across the human proteome and the correlation to protein function. Pacific Symposium on Biocomputing. 2002;**7**:310-322

[27] Steentoft C, Vakhrushev SY, Joshi HJ, Kong Y, Vester-Christensen MB, Schjoldager KT, et al. Precision mapping of the human O-GalNAc glycoproteome through Simple Cell technology. The EMBO Journal. 2013;**32**:1478-1488

Chapter 5

Applications of Mass Spectrometry to the Analysis of Adulterated Food

Gunawan Witjaksono and Sagir Alva

Abstract

Food quality and safety are the major issues in food industry around the world. With the abundance of processed food with long supply chain in the market, food fraud is always a concern. Food fraud is defined as modification of an actual labeling of food chemicals in which expensive, less accessible original ingredients are replaced by lower cost and more accessible alternatives, which is also known as food adulteration. Some of these food adulterations might only affect the public mass financially, but some adulteration might affect others more seriously. Various food authentication techniques can be utilized to ensure safety and quality of food products adhering to the standards, such as DNA-based techniques with polymerase chain reaction, vibrational spectroscopy, electronic nose, and mass spectrophotometry, which has been used widely to estimate pharmaceutical and biological samples. However, most of these techniques still require substantial sample preparation or some have very high sensitivity to adulterants and are prone to give undefined results. Complex mixtures of food adulterants can be identified using very high resolution mass spectroscopy. The chemical compounds and structure of natural and mixtures of the adulterants are examined in this chapter using advanced mass spectroscopy technique and gas chromatography time of flight mass spectroscopy to identify the lard biomarker.

Keywords: mass spectrometry, gas chromatography time-of-flight mass spectrometry, adulterated food, lawful food, mixture food

1. Introduction

Food authentication is a major concern in food industry around the world and significantly affects the global food market. Food fraud as defined in [1] which is alteration of the true labeling of food ingredients by substituting with cheaper and more accessible alternative could affect not only serious consequences to the human health, such as food poisoning and food allergy [2–4] but also loss trust in the confidence of food quality related to the product, company reputation, and religion views [5, 6], which consequently disturbing the global market. Halal and kosher food that are diet intake restrictions are laws for Muslims and Jews religion groups for daily food consumption and have big world market. The global halal market itself worth about $7.049 Billion in 2015, and the analyst projects the market to grow to $1.9 Trillion by 2021 [7]. The continuous growth of such market can only happen when consumers' confidence and trust in the halal labels of the food industry are always maintained and preserved [7–9].

Food fraud as also known as food adulteration is not a recent issue where some of the fraud have been reported earlier, such as adulteration of formula milk with melamine [10–12], mixing of vehicle oil in oil for human consumption in Spain [13], and addition of sawdust to make white bread [10, 14]. The incident of 2008 affected thousands of babies when their milk powders were adulterated [15]. Another incident following that was meat adulteration when prohibited substances were added to the food [16]. The concern of food quality and safety becomes a major priority of both government ministers and the public due to potential financial loss to the state income and increase consumers' health risks that resulted from breaching the food standards.

The food adulteration related to halal and kosher laws is defined as alteration of the original food with pork and its derivatives, such as blood, fat, etc. Lard is a generic ingredient, which is commonly used as a food flavor, mixture, and fat-based blend. Lard has also been reported to be used as an alternative ingredient for adulteration and as a substitute for food-cooking oils, such as butter or margarine. Due to their belief, Muslim communities and Orthodox Jews followers are prohibited to consume lard. Mass spectroscopy (MS) can be used to provide structural details and molecular weight of compounds. Advances of different techniques of MS have emerged significantly. Such advanced techniques utilizing either high resolution mass spectroscopy, that is, GC-MS, or high performance mass spectroscopy, that is, LC-MS, are able to detect more complex compounds with higher accurate identification [17, 18]. Several developments in mass spectrometry for the analysis of the food adulteration have been reported and shown in **Figure 1**.

As shown in **Figure 1**, many food adulterations have been studied in various methods of mass spectroscopy, mainly GC and LC using rapid evaporation ionization spectrometry (REIMS) technique developed initially by Takats et al. [19]. REIMS uses an electrosurgical apparatus that generates surgical smoke after interacting with a solid sample creating ionization and desorption of molecules. Currently, REIMS-based mass spectroscopy has been widely reported for the study of food adulteration, especially for fish and meat adulteration [20, 21]. Another emerging technique is GC-TOF MS mass analyzer for the investigation of a vast number of organic impurities and residues present at the low levels for food quality and safety, surrounding environment, and biological applications [22, 23]. In the analysis of food quality and safety, GC-TOF MS has been utilized to the analysis of in animal-based food origin, such as dioxin-type micro pollutants [24] for the environmental analysis, and GC/GC-TOF MS with negative ionization has been utilized in sediment and fish samples to profile short- and medium-chain

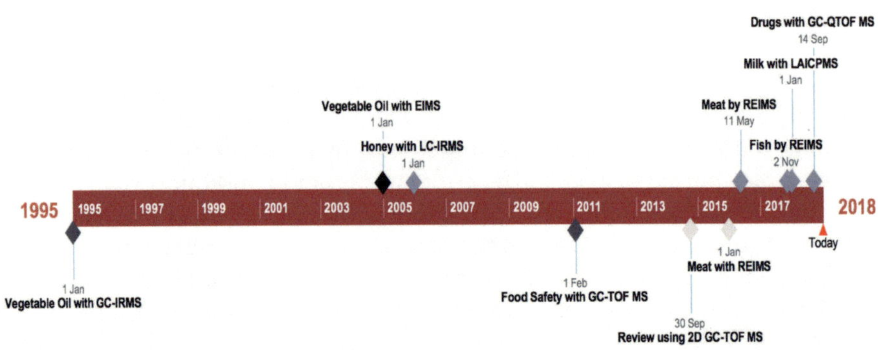

Figure 1.
Some reported work on mass spectrometry development for the investigation of food adulteration.

chlorinated paraffin [25]. In a recent report related to drug-testing investigations, this high-performance mass spectroscopy, that is, GC-TOF MS, is able to analyze doping substances [26].

The purpose of this chapter is to describe advanced mass spectroscopy applications, especially GC-TOF MS technique on investigating food adulteration on pure and mixed meat, covering pig fat, chicken, lab, and cow and to identify the possibility of recognizing a biomarker for lard chemical.

2. Food adulteration

Research communities around the world have been continually working on the food adulteration [27]. Water is a simplest and common food adulterant, especially for milk. Water mixing in milk could degrade the nutritional content, change the taste, and modify the color of milk. Other potentially dangerous adulterants, that is, melamine might be added to replicate natural milk, which seriously increase the health risk [28]. Melamine was used to increase the viscosity of the milk and to keep the composition of fat and carbohydrate to be the same as the original. Such milk adulterant had been reported to cause severe health problems, especially to the infants and young-aged children and created an unusual health outbreak in China in 2008. In some cases, expensive milk is often mixed with the cheaper milk. Reported by Calvano et al. [29], milk from unordinary animals, such as buffalo, camel, and yak, was mixed with ordinary animals, such goat or cow milk. For consumers who are very sensitive to certain types of milk, this kind of food adulteration could trigger in them very serious health problem [30].

2.1 Lawful food

It is a compulsory for most religion followers to follow specific compliances in their daily dietary meals. Such laws, for instance in Islam belief, are to avoid some foods in their dietary consumption, which contain pig meat and its derivatives like ham, bacon, sausages, pork, and lard, except in very rare situations. This requirement is referred to Halal food.

Halal food industry is currently growing significantly and is estimated to reach 20% of global food trade market as world population will consist of 30% Muslim followers by the year of 2025 [31]. Other religions also have defined dietary law; for example, Judaism has kashrut for the Jewish to follow, which also forbids the consumption of pig meat and its by-products [32]. For Hindu religion followers, the consumption of beef and its derivative is not permissible [33]. Many food manufacturers violate the requirement not to practice food fraud that is mostly due to cheap substitute materials.

Muslims and Jews are some of those religious groups that require diet intake restrictions, as they adhere to halal and kosher laws [34, 35], respectively. Although halal and kosher laws have similarities, that is, forbidding consumption of pork and derivatives, blood, etc., they have differences, such as kosher does not forbid alcohol and kosher forbids consumptions of animals that do not chew cud and have cloves, etc. [34]. Although halal and kosher are different, both laws severely forbid the consumption of pork and its derivatives such as lard [34]. Lard is pig fat derived from its adipose tissue and is often used in food production as an emulsion, shortening, or as a substitute to butter, margarine, or cooking oils. The identification of non-halal meat due to lard adulteration is of high significance. Despite many reported work that have been performed to investigate the fingerprint for non-compliance of halal food, such as lard or pig meat [6, 34, 35], the identification of biomarker for non-halal food is still in the early stages.

2.2 Mixed food

A notorious big scandal that hit Europe in 2013 related to food adulteration was the breach of true labeling due to the fraud on the beef sale that has been substituted with horsemeat [36]. The food fraud also occurred in some other part of the world when pharmaceutical preparations and chocolate were suspected to contain traces of pork in 2013 and 2014 in Malaysia [37]. In other countries, like India, it is not uncommon to sell buffalo meat adulterated with other animal meats due to financial issue and availability [38]. Such adulterated meats are very difficult to identify especially when such meats are already in the processed form. The practice of food fraud also occurs on dairy products, for example, butter is mixed with cheaper fats, such as mutton fats, chicken, and pig fats to get higher profits [39]. With these many occurrences of food adulterations around the world, ability to authenticate pure and mixed food has become a crucial aim for everybody.

2.3 Food safety and quality

Food adulteration practices not only destroy consumer trust and confidence in the products and the company reputation but also jeopardize the safety and quality of food consumed. The development of food authentication technique is necessary in food control because of the need of certain compliance in food process and the label to ensure customer confidence and trust to the food product [35, 40]. The authentication technique will also validate the food origin that includes its geographical, gene, and species source, confirming their production processes and their processing techniques [41–43].

The need for food authentication is the result from customer concerns on the food nutrition and their health as well as an assurance of the process control and food quality purposes. Such authentication techniques will also confirm the existence of food adulteration, identify the origin of the food and its ingredients, and improve the food quality and safety for pure and future mixed food.

For this purpose, mass spectroscopy has been very critical in validating and improving food quality and making us caution with any industrial and agriculture chemical to prevent harming our health, disturbing the food supply, and damaging the ecosystem that we depend on for our sustainability. The scientific finding in the environmental, agricultural, and food sciences has been significant to more resourceful and healthier food, improving our quality of life and better living in the world population that is reaching 8 billion and beyond.

3. Food authentication detection

There are several methods that can be used in food authentication process, such as electrophoretic techniques, differential scanning calorimetry (DSC), DNA-based methods (genomics, proteomics), chromatographic methods, isotopic techniques, vibrational and fluorescence spectroscopy, elemental techniques, non-chromatographics mass spectroscopy, sensory analysis, nuclear-magnetic-resonance spectroscopy, immunological techniques together with chemometrics and bioinformatics [40].

DNA-based technique with polymerase chain reaction [38, 44] is a common technique in food authentication testing to ensure halal and kosher brand food products adhere to the standards. However, most of these techniques still require substantial sample preparation or some have very high sensitivity to adulterants and prone to give undefined results if all procedures are not followed exactly.

Research on vibrational spectroscopy-based food authentication techniques is getting more popular [40, 45–52]. This is partly due to the ease of sample preparation with this technique and relatively quick result and non-destructive nature of this method. Such vibrational spectroscopy is able to discriminate with high accuracy. For instance, pork meat and lard in meatball broth [45, 47], imported chocolate [50–53], and vegetable oils [48], etc. are some of the studies. Infrared-based detection techniques, such as FTIR or Fourier transform infrared spectroscopy, are capable of identifying fingerprint of compound molecules when it is incorporated with strong chemometric techniques [47]. Some research findings of lard adulterant are reported either by mixing lard with other animal fats or adulterating lard in food [53–55]. Another work on FTIR spectroscopy by Mansor et al. [56] reported an accuracy up to 100% in performing classification of lard adulterated in virgin coconut oil when the statistical technique, such as discriminant and PLS analysis, is incorporated. However, the limitation of lard detection using FTIR spectroscopy is highlighted in Rohman and Che Man [57] when identifying meat adulteration. Basically, lard has similar IR spectrum with other animal fats and vegetable oils since they are composed with (triacylglycerol) TAG, with different lengths of the fatty acid.

Animal fats have several chemical compositions, which mostly include TAG. In fact, fats share the same fatty acid compounds but different concentrations [58]. According to Rohman and Che Man [57], analysis of fats/oils is possible by focusing on lipid components as fats which is a part biological substance group. Triglyceride is the principal constituent of animal fat, not exception of pig fat. A triglyceride is constructed from three fatty acids and one molecule of glycerol [59]. Lard predominantly consists of saturated fatty acid [59].

Another popular technique that has been continually developed for lard compound detection in food is mass spectroscopy. Several MS methods have been reported, and the important ones are liquid-based chromatography and gas-based chromatography embedded with mass spectrometry (GC-MS and LC-MS).

3.1 Genomics

One of the most popular food authentication methods is the genomics, where verification of foodstuff origin is done by analyzing the cells. Since DNA is similar in the whole somatic cells of a particular species, the original tissue of sample would not affect the results of the test. The advantage of this method is that it can amplify minute samples. Proteomics technique mainly depends on proteins acting as fingerprint of food products and therefore can be applied for a systematic search of new marker proteins. These methods are normally utilized to identify incorrect description and food labeling fraud, that is, detection of meats prohibited by Islamic laws in sausages [35].

3.2 Electronic nose

Electronic nose or e-nose is to replicate human's olfactory technique in identifying a particular substance. E-nose is commonly used metal-oxide gas sensor capable of detecting volatile organic compound (VOC) for variety detection applications including lard adulteration [60] process quality control [61, 62] and used as a formaldehyde sensor [63]. Sensing materials used in the electronic nose for metal oxide sensor are and tungsten trioxide (WO_3) and tin dioxide (SnO_2) because both materials are reported to be very sensitive to many types of volatile compounds.

The sensor selection used in e-nose was based on the chemical compounds found in lard [58]. Decanal was the chemical compound found abundantly in lard

Animal Fat	Decanal (dimensionless)
Lard	17 444
Chicken	586
Beef fat	408.5

Table 1.
Decanal profile, measured in Kovats indices [58].

but did not have significant presence in chicken fat and beef fat. **Table 1** lists the decanal content in the fats of interest in terms of Kovats indices. A set of experiments by Kohl et al. [64] revealed that both the sensing materials used in metal oxide sensors are sensitive to the presence of aldehydes. It is reported here that such sensor is expected to be more sensitive toward lard than other fats.

A scatter plot of sample dataset is shown in **Figure 2** [65]. The dataset consists of nine unique classes of three types of fat each experimented with three different temperatures. Each class consists of 10 observations. Each class is represented in the plot by a unique symbol and an abbreviation where the letter "L" represents a lard sample, "C" represents a chicken fat sample, and "B" a beef fat sample. The numbers 40, 50, and 60 after the letters represent the temperature in degree Celsius. A clear separation can be seen in the plot as except classes "L40" and "C40" where there are no overlaps. The overlap indicates the chemical structure of a chicken fat is very similar to lard, and studies conducted with other techniques have proven that as well.

Figure 3 shows the individual plot of the three classes and their responses at different temperatures [65]. Linear regression lines in the background show an upward trend in sensor response, with lard having the highest gradient out of the three. With the increase of temperatures, the density and rate at which the odor fumes are produced must increase, thus giving rise to a higher sensor response. Besides, this lard has the lowest melting point among the three fats and will therefore melt and turn to gaseous state faster. In terms of settle point values, chicken fat scored the highest above the two as more evident from **Figure 3**. However, the higher settle point values of chicken fat can be explained by the fact that chicken fat melting points are very close to that of lard.

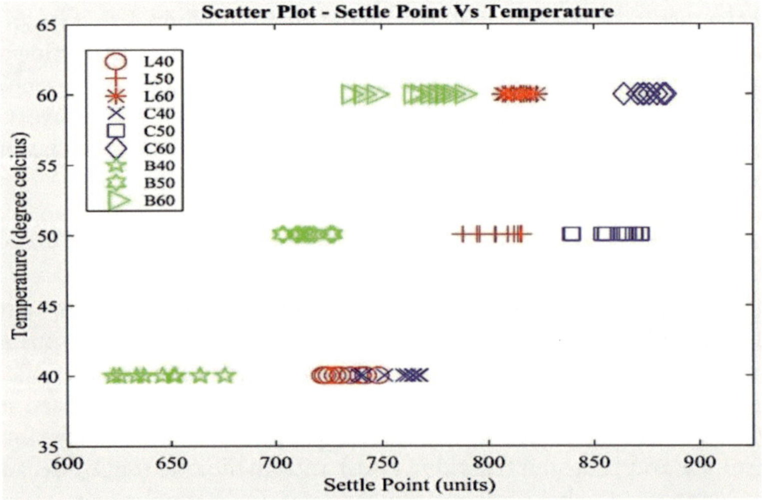

Figure 2.
Scatter plot of the entire dataset [65].

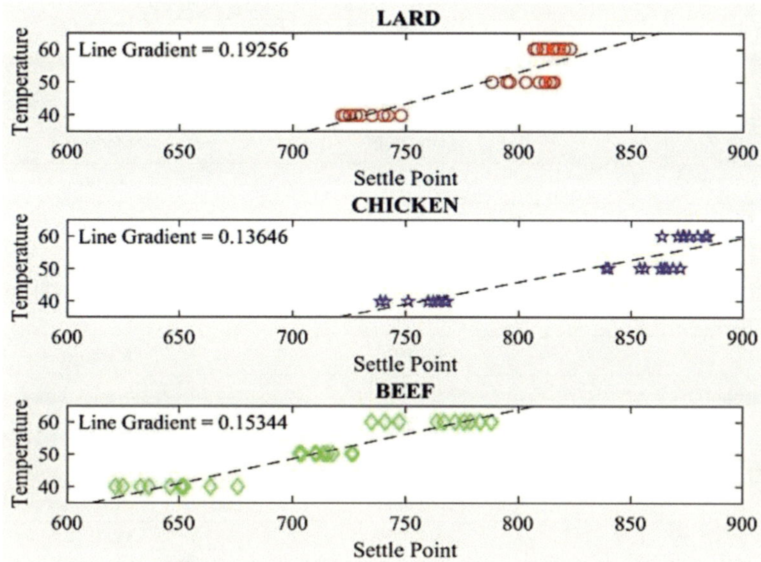

Figure 3.
Response toward change in temperature [65].

3.3 Vibrational spectroscopy

The principle of vibrational spectroscopy follows the concept that atoms in the chemical bonding within the molecule vibrate with certain frequency when it is excited. Such vibration frequency can be explained by the laws of physics and is shown in reported calculation [66]. The calculation of the lowest fundamental frequency of any two atoms that are connected by a chemical bond can be performed by assuming that the bond energy results from the vibration of diatomic harmonic oscillator and follows Hooke's Law according to Eq. (1)

$$v = \frac{1}{2\pi}\sqrt{\frac{k}{\mu}} \qquad (1)$$

where, the vibrational frequency *is* v, the classical force constant is k, and the reduced mass of the two atoms is μ. In contrast to classical spring model for molecular vibration, no continuum of energy levels exists. Instead, there are levels of discrete energy that can be explained by quantum theory. Using the vibrational Hamiltonian, the time-independent Schrödinger equation can be solved for a diatomic molecule. A reduced equation of these levels can be written for the energy levels of diatomic molecules as:

$$E_v = \frac{\left(v + \frac{1}{2}\right)\hbar}{2\pi}\sqrt{\frac{k}{\mu}} \quad (v = 0, 1, 2, \ldots.) \qquad (2)$$

or by using hv as the quantum term, the equation can be reduced to

$$E_v = \left(v + \frac{1}{2}\right)\hbar \quad (v = 0, 1, 2, \ldots.) \qquad (3)$$

At certain extension of the stretch, the bond could eventually breakdown when the vibrational energy goes beyond the dissociation energy. **Table 2** shows the different stretching frequencies. When a fast and objective analysis is required,

	Wavenumber (cm^{-1})	**Intensity**
C≡N	2260–2220	Medium
C≡C	2260–2100	Medium to weak
C=C	1680–1600	Medium
C=N	1650–1550	Medium
	~1600 and ~1500–1430	Strong to weak
C=O	1780–1650	Strong
C−O	1250–1050	Strong
C−N	1230–1020	Medium
O−H (alcohol)	3650–3200	Strong, broad
O−H (carboxylic acid)	3300–2500	Strong, very broad
N−H	3500–3300	Medium, broad
N−H	3300–2700	Medium

Table 2.
Important IR stretching frequencies [67].

fluorescence and absorption spectroscopies in the range of visible to infrared region are better choice. The vibrational spectroscopy is able to provide a fingerprint of the vibrational levels of molecules in the mid-infrared (MIR) radiation (4000–400 cm^{-1}). One of the most common IR spectroscopy techniques is the Fourier transform infrared (FTIR) spectroscopy. FTIR spectroscopy utilizes the use of mid infrared spectroscopy (4000–400 cm^{-1}), which includes the fingerprint region.

3.3.1 Meat sample preparation

All meat samples were collected from a local slaughterhouse and were washed by distilled water. After that, the meat was cut by knife in pieces in the size of 1 cm^2 and stored at −20°C until it was being used. The animal fats extracted from beef, mutton, and chicken body fat as well as lard were collected by rendering the adipose tissues following the method reported by Che Man et al. [53] with little variation.

3.3.2 Post-processing analysis

Data post-processing was done using two software: Spectrograph 1.1 and MATLAB R2017b. Extracting information from spectrum results was carried out using Spectrograph 1.1, where the data are preprocessed as needed. MATLAB R2017b was used to further analyze the results from preprocessing. Principal component analysis (PCA) technique was used to analyze the quality of lard adulteration, while PLS technique was used to analyze the quantity of lard adulteration.

Figure 4 shows FTIR spectra of pure fats. These spectra consist of four regions: 1st region ranging from 4000 to 2500 cm^{-1}, 2nd region ranging from 2500 to 2000 cm^{-1}, 3rd region ranging from 2000 to 1500 cm^{-1}, and lastly the fingerprint region ranging from 1500 to 800 cm^{-1}.

3.4 Mass spectroscopy

The mass spectroscopy methods are fast becoming popular [50, 68]. This method produces unique chemical fingerprinting that can discriminate or verify

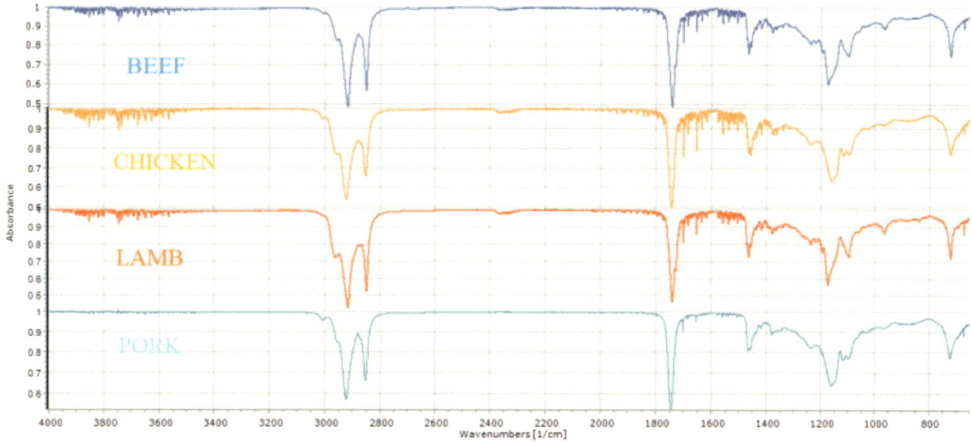

Figure 4.
Spectrogram from FTIR covering 4000–800 cm^{-1}.

foods. MS offers many advantages, such as the identification of mass spectral signal pattern and possible characterization of specific compounds coming from food adulterants. Additionally, MS does not easily react with water, which is different case for vibrational spectroscopy. MS can also provide the plant origin by measuring the specific chemical compounds. However, MS has disadvantages of direct contact requirement to the sample material and larger instrumentation. The spectral resolution of MS is more detail so it has higher possibility of finding fingerprint of food chemicals. MS also gives a higher versatility because of exchangeability of its ion sources. With different ion sources, MS can provide various ionization and is able to perform measurement of chemically different chemical compounds.

Figure 5 shows the spectrogram of different milks using electrospray ionization mass spectroscopy [69]. Obvious differences can be observed among the three milk, (a) cow milk, (b) goat milk, and (c) soy milk, by observing the number of peaks and peak intensities. The three pure milks and the two mixtures score plots are

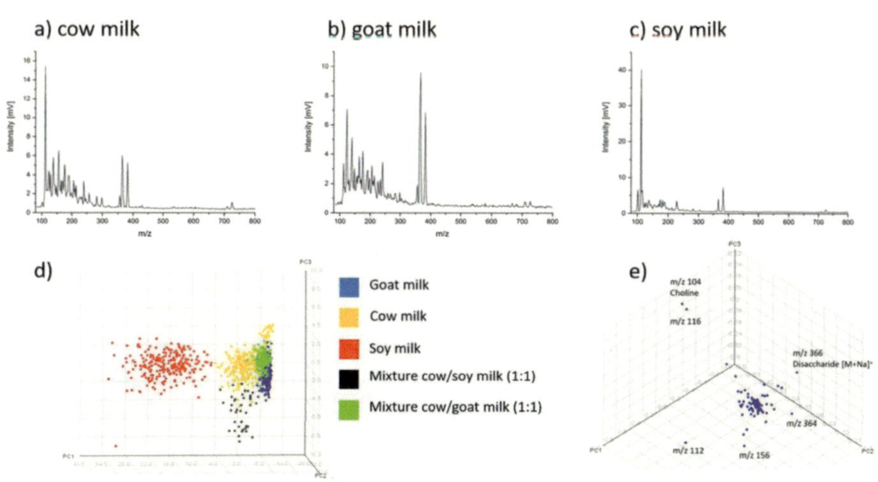

Figure 5.
ESI mass spectroscopy using ESI spectrogram of the three milk, namely cow milk (a) goat milk, (b) soy milk, (c) Plot and (d) shows PCA score plot of the milk data set from the sum of 20 spectra. Each point in plots shows a mass spectrum of goat milk (blue), cow milk (yellow), and coy milk (red), the black-color data show 1:1 mixture of cow and soy milk, and the green shows a mixture of cow and goat milk (e) PCA loading plot of the milk data set. [69].

shown in **Figure 5d**. The spectrograms of the pure milk samples are well separated in the plot, while data points for the mixture of cow and goat milk are positioned in the close proximity of those two types. The data points of cow/soy milk mixture are shown near around the data points of cow milk.

4. Advanced mass spectroscopy

The recent advanced mass spectroscopy instruments offer higher speed, better resolution, higher mass accuracy, and more sensitivity to provide comprehensive qualitative investigation, rapid profiling, and better accuracy detection and quantification of chemical compounds in complex matrices. Thus, such advanced mass spectrometries such as gas chromatography-mass spectrometry (GC-MS) or liquid chromatography-mass spectrometry (LC-MS) are able to investigate and analyze the complex adulterants. These advanced mass spectroscopies operate in scan mode at better spectrum resolution and accurate mass (HRAM).

This improved high-resolution mass spectroscopy is capable in identifying the chemical compounds and mass structure of pure and adulterated processed food, the presence of adulterants that create problems affecting food safety and quality, and the existence of natural toxin, food degradation and contaminations.

4.1 GC-MS

Gas chromatography (GC) configured with electron capture, flame photometric detection, and nitrogen-phosphorous has been used since the early 1970s for residue analysis. The confirmation of results was done with additional use of gas chromatography equipped with a different type of column or detector. Nowadays, using GC integrated with MS, it is able to simultaneously determine and confirm the chemical residues with only one instrument in one analytical run.

Following the commercial of gas chromatography (GC) 50 years ago [70], GC has been used widely in the application involving food adulterant analysis and to perform both quantitative and qualitative analysis of food ingredients, food additives, food adulterants, and contaminants in order to discover nutritional contents, improve food safety, and introduce different food varieties. Furthermore, GC has been reported to be able to identify many organic contaminants at trace levels in complex chemical compounds of food and environmental samples.

Nowadays, gas chromatography integrated to mass spectrometry (GC-MS, GC-HRMS) utilized electron impact ionization (EI) is the most often employed in GC-based MS technique for multi residue chemical analysis in food analysis because of its high selectivity and sensitivity and its ability to screen many pesticides from different chemical compound classes in very complicated matrices in a single run [71]. Advantages of electron impact ionization mass spectroscopy are insignificant influence of molecular structure on response and vast number of characteristic fragments. GC-MS is suitable for analysis of volatile chemicals. Meanwhile, the analysis with more polar compound, LC-MS is more suitable. With the absent of chemical derivatization, GC is commonly used for the analysis of sterols, low chain fatty acids, oils, aroma components and off-flavors, and many contaminants, such as toxins, industrial pollutants, and specific of drugs in foods.

4.2 LC-MS

Liquid chromatography-mass spectrometry (LC-MS) is a combined analytical chemistry technique that separates mixtures with multiple components and provides structural identity of the individual components with high molecular specificity and detection sensitivity. Methods based on liquid chromatography (LC) were applied later after GC, because traditional UV, diode array, and fluorescence detectors are often less selective and sensitive than GC instruments. But in the last few years, the commercial availability of atmospheric pressure ionization caused a dramatic change. Compared to traditional detectors, electrospray (ESI) or atmospheric pressure chemical ionization (APCI) in combination with MS instruments has increased the sensitivity of LC detection by several orders of magnitude.

An analytical methodology using liquid chromatography-mass spectrometry has been reported by Guijarro-Dıez et al. [72] for the detection of the adulteration of saffron samples with gardenia through the determination of geniposide as adulteration marker. **Figure 6** shows the MS spectra obtained for geniposide, and different MS fragments and adducts (Na^+ and $NH4^+$) were obtained for geniposide under ESI^+, whereas when the ESI^- mode was employed, the most abundant ion corresponded to the adduct $[M + HCOO]^-$ (433.1384 *m/z*), and no fragmentation was observed [72].

4.3 High resolution-mass spectroscopy

The instrument of high-resolution mass spectrometry (HRMS) provides better accuracy for the analysis of food adulteration. However, due to high instrumental complexity, HRMS has previously been limited to the most critical applications, such

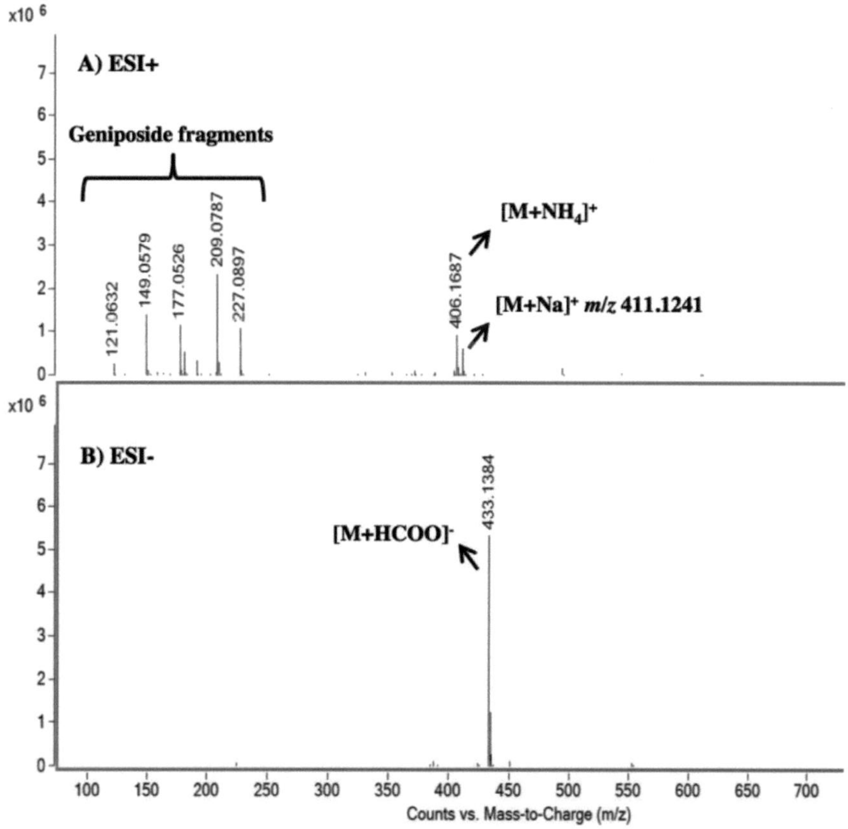

Figure 6.
Mass spectrogram geniposide spectrogram from gardenia extract investigated by LC-MS with (A) ESI^+ and (B) ESI^- [72].

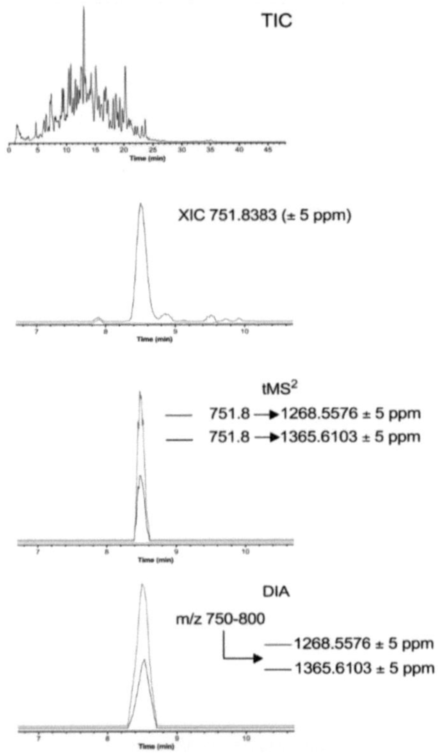

Figure 7.
The observed chromatograms (XICs) for certain signature of myoglobin proteotypic peptide-fragment pairs. The spike was noticed in the chromatograms from beef samples with 1% horse meat (blue indicates extracted blank chromatograms) [73].

as the investigation of natural organic compounds or dioxin-related chemical compounds. The existence of modern HRMS instruments such as time-of-flight (TOF) and Orbitrap instruments has significantly changed the utilization of the equipment. Therefore, high-resolution mass spectrometry (HRMS) has gotten wider acceptance in the last decades for adulterant and residue analysis in food. This positive development is because of the availability of more versatile, robust, sensitive, and advanced instrumentation. The advantages by HRMS compared to classical unit-mass-resolution are ability to provide full-scan spectra, which offers more detail and insight into the mass composition of any sample. As a result, the analyst can measure chemical compounds without the necessity of compound-specific tuning, the need of retrospective data analysis, and has a capability performing an analysis of structural elucidations of suspected chemical compounds. HRMS is still preferable compared with classical hyphenated mass spectrometry in the investigation of quantitative multi residue methods (e.g., pesticides and veterinary drugs). It is one of the most powerful tools for identifying the unknown and non-targeted samples. Improvement of the hardware and software still needs to be addressed by the equipment manufacturers for it to be superior compared to hyphenated mass spectrometry and to be a standard trace analysis tool.

HRMS technology provides proteomic research to facilitate new discovery. The recent HRMS instruments already have the sensitivity, speed, accuracy, and selectivity to deliver comprehensive qualitative analysis, rapid chemical profiling, and high-accuracy analysis and detection of proteins in complex compounds. With these advantages, HRMS-based method was suitable specifically to perform the investigation of meat speciation and to detect food adulteration [73] and is capable to identify quite specific tryptic peptides from targeted proteins.

Motivated by European scandal [74] in which the horse and pig DNA were detected in beef products sold from several retailers, HRMS method developed by Orduna et al. [73] were tested by mixing horse meat in beef meat at concentration 1% w/w. **Figure 7** shows the detection of adulteration of horse proteotypic myoglobin peptide using three different techniques of MS (140,000 FWHM), tMS or DIA [73].

5. GC-TOF MS

GC-TOF MS instrument has two operation modes, in which one mode offers very high scan rates, allowing the segregation of overlapping spectrum peaks by automatically performing deconvolution mass spectral of overlapping spectrum signals [75]. Another type of GC-TOF MS instruments provides high mass resolution, performing data evaluation with a restricted mass window of 0.02 Da [76]. For ion separation GC-TOF MS, single-quad instruments are frequently utilized used. GC-MS systems with quadrupole ion traps integrated with time-of-flight (TOF) mass spectrometers or tandem mass spectrometers are used for the analysis of pure and mixture food.

5.1 Sample preparation

The work by Witjaksono et al. [77] was conducted for total nine meat samples of three different animal meats, that is, chicken, cow, and pig. Each animal meat type is prepared to provide three different samples. The preparation of the animal meat samples and the extraction process of these animal body fats have been done using similar method mentioned before in the FTIR measurements. After obtaining the pure fats, each animal fat (approximately of 50 mg) was dissolved in 0.8 mL hexane. Later, the mixture was stirred for 1 min using an apparatus of vortex mixer and then stored in the dark at −18°C before going to GC-TOF MS analysis.

5.2 GC-TOF MS results

The analysis for this food adulteration was based on GC-TOF MS to identify and study their complex chemical compounds. The equipment used is an Agilent 7693 B GC integrated with TOF MS with hp-5ms column. The analysis was performed for all nine samples, consisting of three samples each from cow, lard, and chicken fats to investigate their aromatic hydrocarbons. The result suggests that the concentration of 1,2,3-trimethyl-benzene, indane, and undecane in lard fat are higher by 250, 14.5, and 1.28 times than chicken fat's concentrations, respectively, and higher by 91.4, 2.3, and 1.24 times higher than cow fat's concentrations, respectively. This initial result promises the possibility of finding biomarkers for non-halal food adulterants.

Table 3 provides the obtained average area covered by each hydrocarbon that is coming from three samples to represent the composition weightage for the different fat types. From **Table 1**, it is obvious that lard is distinctive from the other animal fats in several hydrocarbon compositions. Here are the resulted hydrocarbons that give bigger percentage area in lard in comparison with the other fats: benzene, 1,2,3-trimethyl-; benzene, 1-methyl-3-(1-methylethyl)-; benzene, 1-methyl-4-propyl-; hexanedioic acid, bis(2-ethylhexyl)ester; p-cymene; tridecane; undecane. By using chemometric and bioinformatics analysis techniques, these results could be further analyzed to differentiate and separate the lard fat from the other animal fats.

Hydrocarbon compound	Area %		
	Lard	Chicken fat	Cow fat
2,4-Imidazolidinedione, 5-[3,4-bis[(trimethylsilyl)oxy]phenyl]-3-methyl-5-phenyl-1-(trimethylsilyl)-	0.08299	0.04853	0.141035
Benzene, 1,2,3-trimethyl-	21.33433	0.085155	0.233378
Benzene, 1-methyl-3-(1-methylethyl)-	0.597023	0	0
Benzene, 1-methyl-4-(1-methylpropyl)-	0.013952	0	0.018403
Benzene, 1-methyl-4-propyl-	0.787343	0	0
Benzene, 2-ethyl-1,4-dimethyl-	0.374043	0.432713	0.068181
Decane	21.33433	0	21.167
Decane, 4-methyl-	1.286363	0	0.95659
Hexanedioic acid, bis(2-ethylhexyl) ester	0.583767	0	0
Indane	0.125046	0.008597	0.054052
Naphthalene, 1,2,3,4-tetrahydro-2-methyl-	0.055843	0	0
Nonane, 2-methyl-	0.098341	0	0
Octane, 2,3,7-trimethyl-	0.037965	0	0
p-Cymene	0.447551	0.116017	0
Tridecane	0.22617	0	0
Undecane	10.3596	8.062533	8.382467
Benzene, (2-methyloctyl)-	0	2.472467	2.657267
Benzene, 1-ethyl-4-methyl-	0	0.035777	0.102822
Benzene, 1-methyl-3-propyl-	0	0.680724	0
Cyclohexane, butyl-	0	0.529553	0
Naphthalene, 1,2,3,4-tetrahydro-	0	0.06464	0.138087
o-Cymene	0	0.121686	0
1-Dodecanol, 3,7,11-trimethyl-	0	0	0.222381
Benzene, 1-ethyl-2,4-dimethyl-	0	0	0.062808
Benzene, 1-ethyl-3-methyl-	0	0	0.110763
Heptacosane	0	0	0.056515
Nonane, 2-methyl-	0	0	0.062937
Squalene	0	0	0.342603
Benzene, 4-ethyl-1,2-dimethyl-	0	0.541307	0

Table 3.
Resulted composition of aromatic hydrocarbons for lard, chicken, and cow fats [77].

6. Conclusion

This chapter demonstrated the identification of lard discrimination using GC-TOF MS for cow and chicken fats. GC-TOF MS provides confirmation of lard biomarker that is different with other animal fats for their volatile hydrocarbon compounds in which complex compounds such as benzene, 1-methyl-3-(1-methylethyl)-, hexanedioic acid, bis(2-ethylhexyl)ester, and p-cymene give significant higher compositional percentage in lard fat compared to other animal fats.

Acknowledgements

The author acknowledges the financial support by Universiti Teknologi PETRONAS (UTP) under STIRF-UTP, fund/project code: 0153AA-F71 and International Grant under Universiti Teknologi PETRONAS—Universitas Mercu Buana (UTP-UMB) Collaboration.

Author details

Gunawan Witjaksono[1*] and Sagir Alva[2]

1 Electrical and Electronic Engineering Department, Universiti Teknologi PETRONAS, Perak Darul Ridzuan, Malaysia

2 Mechanical Engineering Department, Universitas Mercu Buana, Jakarta, Indonesia

*Address all correspondence to: gunawan.witjaksono@utp.edu.my

References

[1] Spink J, Moyer DC. Defining the public health threat of food fraud. Journal of Food Science. 2011;**76**:R157-R163

[2] Guardian T. Allergic Teenager's Death After Eating Kebab was Accidental, Rules Coroner. The Guardian. 2017. Available from: https://www.theguardian.com/uk-news/2017/jun/16/teenager-with-dairy-allergy-died-accidentally-rules-coroner

[3] Li DK. Toddler Allergic to Dairy Dies After Pre-School Serves him Grilled Cheese; New York Post; 2017

[4] Barlass T. Child Aged 10 Dies After Drinking Coconut Drink As Importer Admits Label Charges; The Sydney Morning Herald; 2015

[5] FSA. Timeline on Horse Meat Issue. 2013. Available from: http://webarchive.nationalarchives.gov.uk/20150403184406/http://www.food.gov.uk/enforcement/monitoring/horse-meat/timeline-horsemeat [Accessed: February 7, 2018]

[6] Rohman A, Che Man YB. Analysis of pig derivatives for halal authentication studies. Food Reviews International. 2012;**28**:97-112

[7] Reuters T. State of the Global Islamic Economy Report 2016/17; 2016

[8] Barnett J, Begen F, Howes S, Regan A, McConnon A, Marcu A, et al. Consumers' confidence, reflections and response strategies following the horsemeat incident. Food Control. 2016;**59**:721-730

[9] Schmutzler M, Beganovic A, Böhler G, Huck CW. Methods for detection of pork adulteration in veal product based on FT-NIR spectroscopy for laboratory, industrial and on-site analysis. Food Control. 2015;**57**:258-267

[10] Tähkäpää S, Maijala R, Korkeala H, Nevas M. Patterns of food frauds and adulterations reported in the EU rapid alert system for food and feed and in Finland. Food Control. 2015;**47**:175-184

[11] Guan N, Fan Q, Ding J, Zhao Y, Lu J, Ai Y, et al. Melamine-contaminated powdered formula and urolithiasis in young children. New England Journal of Medicine. 2009;**360**:1067-1074

[12] Jia C, Jukes D. The national food safety control system of China—A systematic review. Food Control. 2013;**32**:236-245

[13] Abaitua Borda I, Philen RM, Posada de la Paz M, Gomez de la Camara A, Diez Ruiz-Navarro M, Gimenez Ribota O, et al. Toxic oil syndrome mortality: The first 13 years. International Journal of Epidemiology. 1998;**27**:1057-1063

[14] Wood R. Symposium on Food Identification and Authentication; 2012

[15] Gu Y, Han W, Zheng L, Jin B. Using IoT technologies to resolve the food safety problem—An analysis based on Chinese food standards. In: Wang FL, Lei J, Gong Z, Luo X, editors. Web Information Systems and Mining: International Conference, WISM; 26-28 October, 2012; Chengdu, China. Proceedings. Berlin, Heidelberg: Springer Berlin Heidelberg; 2012. pp. 380-392

[16] Liu Y, Han W, Zhang Y, Li L, Wang J, Zheng L. An internet-of-things solution for food safety and quality control: A pilot project in China. Journal of Industrial Information Integration. 2016;**3**:1-7

[17] McMaster MC. GC/MS: A Practical User's Guide. 2nd ed. Hoboken, New Jersey: John Wiley and Sons; 2008. ISBN: 978-0470101636

[18] McMaster MC. LC/MS: A Practical User's Guide. Hoboken, New Jersey: John Wiley and Sons; 2005. ISBN: 978-0471655312

[19] Takats Z, Denes J, Kinross J. Identifying the margin: A new method to distinguish between cancerous and noncancerous tissue during surgery. Future Oncology. 2012;**8**(2):113-116. DOI: 10.2217/fon.11.151

[20] Balog J, Perenyi D, Guallar-Hoyas C, Egri A, Pringle SD, Stead S, et al. Identification of the species of origin for meat products by rapid evaporative ionization mass spectrometry. Journal of Agricultural and Food Chemistry. 2016;**64**(23):4793-4800. DOI: 10.1021/acs.jafc.6b01041

[21] Black C, Chevallier OP, Haughey SA, Balog J, Stead S, Pringle SD, et al. A real time metabolomic profiling approach to detecting fish fraud using rapid evaporative ionisation mass spectrometry. Metabolomics. 2017;(12):1-13. DOI: 10.1007/s11306-017-1291-y

[22] Hernández F, Portolés T, Pitarch E, López FJ. Gas chromatography coupled to high-resolution time-of-flight mass spectrometry to analyze trace-level organic compounds in the environment, food safety and toxicology. Trends in Analytical Chemistry. 2011;**30**:388-400

[23] Tranchida PQ, Franchina FA, Dugo P, Mondello L. Comperhensive two-dimensional gas chromatography-mass spectrometry: Recent evolution and current trends. Mass Spectrometry Reviews. 2014;**35**:524-534

[24] Planche C, Ratel J, Mercier F, Blinet P, Debrauwer L, Engel E. Assessment of comprehensive two-dimensional gas chromatography-time-of-flight mass spectrometry based methods for investigating 206 dioxin-like micropollutants in animal-derived food matrices. Journal of Chromatography. A. 2015;**1392**:74-81

[25] Xia D, Gao L, Zheng M, Tian Q, Huang H, Qiao L. A novel method for profiling and quantifying short- and medium-chain chlorinated paraffins in environmental samples using comprehensive two-dimensional gas chromatography-electron capture negative ionization high-resolution time-of-flight mass spectrometry. Environmental Science & Technology. 2016;**50**:7601-7609

[26] Abushareeda W, Tienstra M, Lommen A, Blokland M, Sterk S, Kraiem S, et al. Comparison of gas chromatography quadrupole time-of-flight and quadrupole orbitrap mass spectrometry in anti-doping analysis: I. Detection of anabolic-androgenic steroids. 2018;**32**(23):2055-2064. DOI: 10.1002/rcm.8281

[27] Downey G. Advances in Food Authenticity Testing. 1st ed. UK: Elsevier; 2016

[28] Handford CE, Campbell K, Elliott CT. Impact of milk fraud on food safety and nutrition with special emphasis on developing countries. Comprehensive Reviews in Food Science and Food Safety. 2016;**15**:130-142

[29] Calvano CD, De Ceglie C, Monopoli A, Zambonin CG. Detection of sheep and goat milk adulterations by direct MALDITOF MS analysis of milk tryptic digests. Journal of Mass Spectrometry. 2012;**47**:1141-1149

[30] Sampson HA. Food allergy. Journal of Allergy and Clinical Immunology. 2003;**111**:S540-S547

[31] Abdullah Amqizal HI, Al-Kahtani HA, Ismail EA, Hayat K, Jaswir I. Identification and verification of porcine DNA in commercial gelatin and gelatin containing processed foods. Food Control. 2017;**78**:297-303

[32] Schröder MJA. Food Quality and Consumer Value: Delivering Food that Satisfies; 2003

[33] Bonne K, Verbeke W. Religious values informing halal meat production and the control and delivery of halal credence quality. Agriculture and Human Values. Berlin, Heidelberg: Springer; 2008;**25**:35-47

[34] Regenstein JM, Chaudry MM, Regenstein CE. The kosher and halal food Laws. Comprehensive Reviews in Food Science and Food Safety. 2003;**2**:111-127

[35] Chuah L-O, He XB, Effarizah ME, Syahariza ZA, Shamila-Syuhada AK, Rusul G. Mislabelling of beef and poultry products sold in Malaysia. Food Control. 2016;**62**:157-164

[36] FSA. Timeline on Horse Meat Issue. 2013. Available: http://webarchive.nationalarchives.gov.uk/20150403184406/http://www.food.gov.uk/enforcement/monitoring/horse-meat/timeline-horsemeat [Accessed: February 7, 2018]

[37] Flaudrops C, Armstrong N, Raoult D, Chabrière E. Determination of the animal origin of meat and gelatin by MALDITOF-MS. Journal of Food Composition and Analysis. 2015;**41**:104-112

[38] Mane BG, Mendiratta SK, Tiwari AK, Bhilegaokar KN. Development and evaluation of polymerase chain reaction assay for identification of buffalo meat. Food Analytical Methods. 2012;**5**:296-300

[39] Nurrulhidayah AF, Arieff SR, Rohman A, Amin I, Shuhaimi M, Khatib A. Detection of butter adulteration with lard using differential scanning calorimetry. International Food Research Journal. 2015;**22**:832-839

[40] Danezis GP, Tsagkaris AS, Camin F, Brusic V, Georgiou CA. Food authentication: Techniques, trends & emerging approaches. TrAC Trends in Analytical Chemistry. 2016;**85**:123-132

[41] Alamprese C, Amigo JM, Casiraghi E, Engelsen SB. Identification and quantification of Turkey meat adulteration in fresh, frozen-thawed and cooked minced beef by FT-NIR spectroscopy and chemometrics. Meat Science. 2016;**121**:175-181

[42] Barbin DF, Sun D-W, Su C. NIR hyperspectral imaging as non-destructive evaluation tool for the recognition of fresh and frozen-thawed porcine longissimus dorsi muscles. Innovative Food Science & Emerging Technologies. 2013;**18**:226-236

[43] Morsy N, Sun D-W. Robust linear and non-linear models of NIR spectroscopy for detection and quantification of adulterants in fresh and frozen-thawed minced beef. Meat Science. 2013;**93**:292-302

[44] Vlachos A, Arvanitoyannis IS, Tserkezou P. An updated review of meat authenticity methods and applications. Critical Reviews in Food Science and Nutrition. 2016;**56**:1061-1096

[45] Kurniawati E, Rohman A, Triyana K. Analysis of lard in meatball broth using Fourier transform infrared spectroscopy and chemometrics. Meat Science. 2014;**96**:94-98

[46] Meza-Márquez OG, Gallardo-Velázquez T, Osorio-Revilla G. Application of mid-infrared spectroscopy with multivariate analysis and soft independent modeling of class analogies (SIMCA) for the detection of adulterants in minced beef. Meat Science. 2010;**86**:511-519

[47] Rahmania H, Sudjadi, Rohman A. The employment of FTIR spectroscopy in combination with chemometrics for analysis of rat meat in meatball formulation. Meat Science. 2015;**100**:301-305

[48] Rohman A, Che Man YB, Hashim P, Ismail A. FTIR spectroscopy combined

with chemometrics for analysis of lard adulteration in some vegetable oils Espectroscopia FTIR combinada con quimiometría Para el análisis de adulteración con grasa de cerdo de aceites vegetales. CyTA Journal of Food. 2011;**9**:96-101

[49] Rohman A, Che Man YB. FTIR spectroscopy combined with chemometrics for analysis of lard in the mixtures with body fats of lamb, cow and chicke. International Food Research Journal. 2010;**17**:519-527

[50] Suparman WS, Sundhani E, Saputri SD. The use of Fourier transform infrared spectroscopy (FTIR) and gas chromatography mass spectroscopy (GCMS) for halal authentication in imported chocolate with various variants. Analysis. 2015;**2**:03

[51] Xu L, Cai CB, Cui HF, Ye ZH, Yu XP. Rapid discrimination of pork in halal and non-halal Chinese ham sausages by Fourier transform infrared (FTIR) spectroscopy and chemometrics. Meat Science. 2012;**92**:506-510

[52] Yang H, Irudayaraj J, Paradkar MM. Discriminant analysis of edible oils and fats by FTIR, FT-NIR and FT-Raman spectroscopy. Food Chemistry. 2005;**93**:25-32

[53] Che Man YB, Syahariza ZA, Mirghani MES, Jinap S, Bakar J. Analysis of potential lard adulteration in chocolate and chocolate products using Fourier transform infrared spectroscopy. Food Chemistry. 2005;**90**:815-819

[54] Rohman A, Erwanto Y, Man YBC. Analysis of pork adulteration in beef meatball using Fourier transform infrared (FTIR) spectroscopy. Meat Science. 2011;**88**(1):91-95

[55] Syahariza Z, Che Man YB, Selamat J, Bakar J. Detection of lard adulteration in cake formulation by Fourier transform infrared (FTIR) spectroscopy. Food Chemistry. 2005;**92**(2):365-371

[56] Mansor TST, Che Man YB, Rohman A. Application of fast gas chromatography and Fourier transform infrared spectroscopy for analysis of lard adulteration in virgin coconut oil. Food Analytical Methods. 2011;**4**:365-372

[57] Rohman A, Che Man YB. Quantification and classification of corn and sunflower oils as adulterants in olive oil using chemometrics and FTIR spectra. The Scientific World Journal. 2012;**2012**:250795

[58] Nurjuliana M, Che Man YB, Hashim DM. Analysis of lard's aroma by an electronic nose for rapid halal authentication. Journal of the American Oil Chemists' Society. 2011;**88**(8):75-82

[59] Asif M. General chemistry, composition, identification and qualitative tests of fats or oils. Journal of Pharmaceutical Research & Opinion. 2011;**1**(2):52-64

[60] Tian X, Wang J, Cui S. Analysis of pork adulteration in minced mutton using electronic nose of metal oxide sensors. Journal of Food Engineering. 2013;**119**(4):744-749

[61] Barbri N, El Llobet E, El Bari N, Correig X, Bouchikhi B. Electronic nose based on metal oxide semiconductor sensors as an alternative technique for the spoilage classification of red meat. Sensors. 2008:142-156

[62] Tudu B, Metla A, Das B, Bhattacharyya N, Jana A, Ghosh D, et al. Towards versatile electronic nose pattern classifier for black tea quality evaluation: An incremental fuzzy approach. IEEE Transactions on Instrumentation and Measurement. 2009;**58**(9):3069-3078

[63] Xu K, Zeng D, Tian S, Zhang S, Xie C. Hierarchical porous SnO_2 micro-rods topologically transferred from tin oxalate for fast response sensors to trace formaldehyde. Sensors and Actuators B: Chemical. 2014;**190**:585-592

[64] Kohl D, Heinert L, Bock J, Hofmann T, Schieberle P. Systematic studies on responses of metal-oxide sensor surfaces to straight chain alkanes, alcohols, aldehydes, ketones, acids and esters using the SOMMSA approach. Sensors and Actuators B: Chemical. 2000;**70**(1-3):43-50

[65] Latief M, Khorsidtalab A, Saputra I, Akmeliawati R, Nurashikin A, Jaswir A, et al. Rapid lard identification with portable electronic nose. In: IOP Conf. Series: Materials Science and Engineering. Vol. 260. 2017. p. 012043

[66] Burns DA, Ciurczak EW. Handbook of Near-Infrared Analysis. Third ed. London, UK: Pearson; 2007

[67] Bruice PY. Organic Chemistry. 8th ed. London, UK: Pearson; 2016

[68] Ahmad Nizar NN, Nazrim Marikkar JM, Hashim DM. Differentiation of lard, chicken fat, beef fat and mutton fat by GCMS and EA-IRMS techniques. Journal of Oleo Science. 2013;**62**:459-464

[69] Gerbig S, Neese S, Penner A, Spengler B, Schulz S. Real-time food authentication using a miniature mass spectrometer. Analytical Chemistry. 2017;**89**(20):10717-10725. DOI: 10.1021/acs.analchem.7b01689

[70] Lehotay SJ, Hajslova J. Application of gas chromatography in food analysis. Trends in Analytical Chemistry. 2002;**21**(9-10):686-697

[71] Anna Stachniuk A, Emilia Fornal E. Liquid chromatography-mass spectrometry in the analysis of pesticide residues in food. Food Analytical Methods. 2016;**9**:1654-1665. DOI: 10.1007/s12161-015-0342-0

[72] Guijarro-Dıez M, Castro-Puyana M, Crego AL, Marina ML. Detection of saffron adulteration with gardenia extracts through the determination of geniposide by liquid chromatography-mass spectrometry. Journal of Food Composition and Analysis. 2016;**55**: 30-37. DOI: 10.1016/j.jfca.2016.11.004

[73] Orduna AR, Husby E, Yang CT, Ghoshm D, Beaudry F. Detection of meat species adulteration using high-resolution mass spectrometry and a proteogenomics strategy. Food Additives & Contaminants: Part A. 34(7):1110-1120. DOI: 10.1080/19440049.2017.1329951

[74] DG Health and Consumers, European Commission. Horse Meat Issue; DG Health and Consumers. 2013. Brussels, Belgium: European Commission. Available from: http://ec.europa.eu/food/food/horsemeat/tests_results_en.htm [Accessed December 22, 2014]

[75] de Koning S, Lach G, Linkerhagner M, Loscher R, Horst TP, Brinkman UA. Trace-level determination of pesticides in food using difficult matrix introduction-gas chromatography-time-of-flight mass spectrometry. Journal of Chromatography. A. 2003;**1008**:247-252

[76] Cajka T, Hajslova J. Gas chromatography-high-resolution time-of-flight mass spectrometry in pesticide residue analysis: Advantages and limitations. Journal of Chromatography. A. 2004;**1058**:251-261

[77] Witjaksono G, Khir MHM, Saputra I, Mian MU, Rabih AAS, Junaid M, Setiawan LF, Akmeliawati R, Jaswir I, Siddiqui MA. Fourier Transform Infrared Spectroscopy Detection Analysis of Lard in Meat Mixtures. Unpublished

Chapter 6

Molecular Biology of Lung Cancer and Future Perspectives for Screening

Giulio Tarro, Moreno Paolini and Alessandra Rossi

Abstract

Lung cancer patients have the highest mortality among patients with solid tumors worldwide and their prognosis is strictly stage-associated. However, only 15–20% of patients are diagnosed in stage I, since these early tumors are frequently asymptomatic. Early detection of lung cancer, which allows effective therapeutic intervention, is a promising approach to lowering its mortality rate. However, conventional diagnostic methods for lung cancer, such as chest X-ray and CT of the chest, produce high costs and potentially false-positive results. Thus, the discovery of highly sensitive, specific, noninvasive, and cost-effective lung cancer biomarkers combined with conventional approaches, such as X-rays, may improve the sensitivity of lung cancer screening. Herein, we summarize the most recent studies about the molecular pathology of lung cancer and discuss the advancements expected in the near future, including the potential biomarkers and liquid biopsy approaches for the detection of lung cancer in populations at risk of developing this disease.

Keywords: lung cancer, CT, biomarkers, lung cancer screening, liquid biopsy, cfDNA, CTCs

1. Introduction

Despite multimodality treatment strategies including surgery, radiotherapy, chemotherapy, and targeted therapy, lung cancer is still the first leading cause of cancer-related death in the world with the 5-year lung cancer survival rate remaining as low as 15% [1–3]. The most common histologies are summarized as non-small cell lung cancer (NSCLC) and account for 80–85% of newly diagnosed cases. Surgery is the standard of care for functionally operable early stage NSCLC and resectable stage IIIA disease and possesses a potential for cure. However, only 20% of NSCLC are resectable at diagnosis [4]. Histology and cytology of thoracic biopsies are currently the gold standard that asserts early diagnosis of lung cancers detected by thoracic imagery. Nevertheless this approach is costly and often detects false positive nodules that turn out not to be cancers. Therefore, early detection approaches—especially directed toward the population at high risk for the development of this disease—remain an unmet clinical need. Several studies have been performed to define the ideal approach that should be sensitive, specific, reliable, and reproducible for early diagnosis of lung cancer or for prediction of the development of this disease in subjects at risk. This review summarizes the molecular biology of lung cancer and conventional diagnostic methods currently used, with a particular attention on the development of new screening approaches such as liquid biopsy to improve the early detection of this disease.

2. The molecular landscape of lung cancer

2.1 Gene mutations

Recent advances in next-generation sequencing (NGS) and other high-throughput genomic profiling platforms have allowed the examination of the breadth of genetic mutations within lung cancer. The most common mutation is in the Kirsten rat sarcoma (KRAS) oncogene, occurring in approximately 30% of adenocarcinomas (Ade) predominantly in patients with a history of smoking [5]. BRAF is mutated in approximately 3% of patients (with half of cases being the V660E mutation) [6]. Along with KRAS and BRAF, epidermal growth factor receptor (EGFR) mutations were discovered in patients with Ade and small cell lung cancer (SCLC) [5]. Moreover, mutations and amplifications in many oncogenes have been identified, including HER2, MET, as well as fusion oncogenes involving anaplastic lymphoma kinase (ALK), neuregulin 1 (NRG1), neurotrophic tyrosine kinase receptor type 1 (NTRK1), and RET [7–13]. Microtubule-associated protein-like 4 (EML4) and ALK fusion gene is another important driver gene in lung cancer, which was discovered by Soda et al. in 2007 [10]. In NSCLC, EML4/ALK is an aberrant fusion gene that encodes a cytoplasmic chimeric protein with constitutive kinase activity. The incidence of EML4/ALK fusion in cohorts of patients with NSLCL ranges from 1.6% to as high as 19.3%. Genes such as discoidin domain-containing receptor 2 (DDR2); fibroblast growth factor receptor 1, 2, and 3 (FGFR1, FGFR2, FGFR3); and genes in the phosphatidylinositol 3 kinase (PI3K) pathway seem instead to be more commonly mutated in squamous cell carcinoma (SCC). Many of these mutations have been validated by preclinical studies as driver mutations [14–16]. Aberration in stem cell factor receptor tyrosine kinase (c-KIT), PI3K catalytic subunit alpha (PIK3CA), PI3K/AKT/mTOR, phosphatase and tensin homolog (PTEN), insulin-like growth factor receptor (IGFR1), and hedgehog (Shh) signaling pathways have been identified in lung cancer [7] (**Table 1**).

2.2 ALDH family proteins

The cancer stem cell model proposes that tumor progression, drug resistance, metastasis, and relapse after therapy may be driven by a subset of cells within the tumor: the cancer stem cells (CSCs) [17–20]. Recent evidences suggest that like other tumors, human lung cancers may also harbor CSC populations. Human alcohol dehydrogenase (ADH) and aldehyde dehydrogenase (ALDH) are the principal enzymes responsible for ethanol metabolism and have heterogeneous tissue distribution. Isoenzymes of ADH participate in bioamine, prostaglandin, and retinoid acid metabolism [21]. The second enzyme ALDH belongs to a large family of intracellular enzymes that participate in cellular detoxification, differentiation, and drug resistance through the oxidation of endogenous and exogenous aldehydes to carboxylic acids [22]. The ALDH superfamily currently consists of 19 known putatively functional genes in 11 families and 4 subfamilies with distinct chromosomal locations [23–25]. Several studies have explored the biological significance of ALDH in cancers such as head and neck cancer, colon cancer, breast cancer, papillary thyroid carcinoma, and specifically lung cancer, where they have provided supportive evidence for the association between ALDH activity and lung cancer

Altered genes	Histology	Mutation frequency (%)	References
KRAS	Ade	30	[5]
BRAF	NSCLC	3	[6]
EGFR	ADC	19	[5]
HER2	Ade	10	11
MET	NSCLC	8–10	[6]
FGFR1	SCC	22	[14, 15]
CD74-NRG1	Ade	27	[8]
NTRK1	Ade	3.3	[13]
KIF5B- and *CCDC6-RET*	Ade	0.9	[13]
EML4/ALK	NSCLC	6.7	[10]
DDR2	SCC	4	[16]
c-KIT	SCLC, NSCLC	30–40 (SCLC), 40 (NSCLC)	[7]
PIK3CA	SCC	70	[7]
PTEN	SCLC	70	[7]
IGFR1	NSCLC	29.2	[80]

Table 1.
Recurrent somatic genetic alterations detected in lung cancer.

stem cells [26–32]. ALDH1A1 seems to be co-expressed with other NSCLC stem cell markers such as leucine-rich repeat-containing G-protein-coupled receptor 5 (LGR5) in NSCLC tissues, and their expression is significantly associated with stage disease and poor prognosis [33]. It was reported that ALDH1A1-negative expression in lung cancer patients corresponds to shorter survival compared to those with ALDH1A1-positive expression and that ALDH1A1 overexpression was associated with a favorable outcome. Moreover, high expression of ALDH1A1 mRNA was found to be correlated to a better overall survival (OS) in all NSCLC patients followed for 20 years. In addition, high expression of ALDH1A1 mRNA was also found to be correlated to better OS in Ade patients but not in SCC patients. These results strongly support that ALDH1A1 mRNA in NSCLC is associated with better prognosis. However, there are other contradictory results indicating that ALDH1 cytoplasmic expression was associated with poor prognosis in several tumors, such as NSCLC [34]. Jiang et al. also showed that ALDH1A1 expression was positively correlated with the stage and grade of lung tumors and related to a poor prognosis [34]. A recent meta-analysis showed that increased ALDH1A1 expression is associated with poor OS and disease-free survival in lung cancer patients [35]. Previous studies showed that also several other ALDH isoforms are involved in lung cancer as ALDH3A1, highly expressed in two types of NSCLC, Ade and SCC, and ALDH3B1 expression was also found to be upregulated in a high percentage of human tumors, particularly in lung cancer [36–38].

3. Lung cancer screening

Cancer screening is promising for malignancies with a stage-dependent prognosis, and it aims to reduce morbidity and mortality through detection of cancer at an early stage. In general, the screening programs have to be subjected to a rigorous risk-benefit assessment taking into account the endpoints as cancer-related mortality, overall mortality, morbidity, patient-reported outcome, and costs. All the screening programs need a transparent system of quality assurance.

3.1 Low-dose computed tomography

Several studies on lung cancer screening were conducted mainly by using chest X-rays (CXR) for imaging alongside sputum cytology. The National Lung Screening Trial (NLST) enrolled 53,000 individuals aged 55–74 years with a 30-pack-year smoking history, and participants were randomly assigned to radiography or low-dose CT. The low-dose CT group had a 20% reduction in lung cancer mortality and a 6–7% reduction in all-cause mortality [39]. The International Early Lung Cancer Action Program (I-ELCAP) analyzed retrospectively the outcomes of more than 21,000 patients after the completion of the NLST. Different size threshold for nodule diameters resulted in different cancer diagnosis rates. Increasing the threshold from 5 to 0 mm to 6–0, 7–0, 8–0, or 9–0 mm also changed the frequencies of positive results [40]. With respect to North American, European studies performed on a smaller number of individuals at risk of lung cancer showed somewhat inconsistent and less significant results [41–43]. Although these studies showed an improved stage distribution in favor of earlier stages, better resectability of the tumors, and also improved survival, an effect on overall mortality could not be demonstrated [39, 44, 45]. Aside from the morbidity and mortality that is not justified within this context, the expenses turn out to be substantial, as thoracic imagery can be repeated, leading also to debated benefit risks. Despite the progress made in imagery, which allowed the detection of nodules less than 3–4 mm and even the definition of the malignant or benign features, currently cancerous lesions less than 1 mm cannot be detected by imagery [46]. A major drawback of low-dose CT is the large number of false-positive tests and the diagnosis of indolent tumors which in turn lead to an increased morbidity from unnecessary surgical treatment [47–50]. Thus, even if the imagery can allow early stage asymptomatic and operable lung cancer detection, these approaches are not satisfactory because of high cost, high risk of radiation exposure, and poor sensitivity and specificity.

3.2 Biomarkers for lung cancer detection

The discovery of cancer biomarkers, specific molecules that help to distinguish between normal and cancerous conditions, may potentially be used to develop a more effective diagnostic tool for cancer. Body fluids (blood, pleural effusion, etc.) that are in contact with tumors are enriched with proteins shed from cancer cells. Proteins secreted from cancer cells could enter the blood circulation and have the potential to be monitored in plasma/serum. Carcinoembryonic antigen (CEA) is an oncofetal protein not typically expressed in adult tissues. In lung cancer the CEA levels in blood are elevated and are inversely correlated with the response to cancer therapy. Therefore, this marker is used for the detection of cancer recurrence and the prediction of a poor survival rate. CYFRA-21-1 is a fragment of cytokeratin 19 that is typically associated with epithelial cell cancers including NSCLC. This marker is correlated with disease response and the prognosis of cancer but cannot be used to identify cancer patients from patients with respiratory diseases. The sensitivity of CYFRA 21-1 for NSCLC ranges between 23 and 70% [51, 52]. Neuron-specific enolase (NSE) is a glycolysis enzyme produced in neuronal cells and cells with neuroendocrine differentiation. SCLC is of neuroendocrine origin, and therefore NSE is found to be elevated in patients' blood [53]. Tumor M2-pyruvate kinase (PKM2) is a dimeric form of the pyruvate kinase isoenzyme type M2 that is increased in various cancers [52, 54]. C-reactive protein (CRP) is an acute-phase protein, the levels of which rise in response to inflammatory conditions such as lung cancer. However, recent studies suggested that CRP could be used as a prognostic biomarker of lung cancer and angiogenesis [55]. Serological markers such as CEA, NSE, and CYFRA 21-1 are used for the monitoring of treatment effects in lung cancer, but their diagnostic value as screening biomarkers is still being debated [56, 57]. To date, no useful marker has been identified for the screening of asymptomatic patients. Ideally, a biomarker should have a sensitivity and specificity of 100%, a goal that is almost never achieved. One strategy potentially increasing both parameters is to combine several biomarkers into a screening marker panel. Several studies with smaller panels encompassing few markers provided first evidence that simultaneous analysis of several antigens have a higher potential for separating patients with lung cancer from controls [56]. Combined with other noninvasive methods, this may allow for further refinement of lung cancer screening [58].

4. Future perspectives

4.1 New potential lung cancer biomarkers

Proteomics studies showed new lung cancer biomarkers that can be tested in the blood (**Table 2**). Plasma kallikrein (KLKB1) enzyme cleaves Lys-Arg and Arg-Ser bonds in kininogen to release bradykinin and has functions related to blood coagulation. Studies evidenced how serum levels of its fragmentation form were increased in lung cancer samples compared with normal control sera [59, 60]. Serum amyloid A (SAA) proteins are a family of apolipoproteins associated with the high-density lipoprotein (HDL) complex that are secreted during the acute phase of inflammation. In particular, isoforms SAA1/2 were detected in Ade patients' sera but not in healthy donors' sera using liquid chromatography/mass spectrometry (LC-MS/MS). This protein was also detected in tissue [59, 61]. Haptoglobin (Hp) is a free hemoglobin-binding glycoprotein that inhibits the oxidative stress of hemoglobin and assists in hemoglobin uptake. It is a tetramer constituted by two α and two β chains. High levels of Hp have been reported in various cancer types including lung cancer. Proteomics analysis showed Hp β chain peptide levels to be threefold higher in lung cancer patients' sera with respect to control subjects [59, 62]. Complement component 9 (C9) protein, a terminal constituent of the membrane attack complex, plays a role in the immune response by forming plasma membrane pores. This protein was identified in sera of patients with SCC by glycoproteomics approaches. Its protein levels were significantly higher in SCC patients than those in healthy donors and in patients with other cancer types [59, 63]. Insulin-like growth factor-binding protein-2 (IGFBP-2), member of the insulin-like growth factor-binding protein family, inhibits IGF-mediated growth and development rates. Increased levels of IGFBP-2 have been found in solid tumors and in blood from patients with glioma and colorectal, prostate, and breast cancers above all at advanced stage disease. Recently circulating anti-IGFBP-2 autoantibodies and IGFBP-2 combined markers showed increased diagnostic sensitivity and specificity for lung cancer with respect to IGFBP-2 alone [64]. Peroxiredoxin 1 (PRX1) and peroxiredoxin 2 belong to a family of ubiquitous multifunctional antioxidant proteins. The main function of PRX1 is to eliminate peroxides generated during metabolism. PRX1 is also involved in the inhibition of oncogenes, and its protein levels were found to be higher in human cancer cells and tissues. Recently, PRX1 was also identified in lung cancer patients' plasma by mass spectrometry-based screening technology. Plasma PRX1 levels were increased in patients with lung cancer and also in subjects exposed to asbestos [65]. Endoglin (CD105) is a major cell membrane glycoprotein

Protein	Histology	Method	Reference
KLKB1	Lung cancer	LC-MS/MS, WB	[59, 60]
SAA	Ade	MRM, LC-MS/MS	[59, 61]
Hp β chain peptide	Lung cancer	ELISA, WB	[59, 62]
Complement C9	SCC	ELISA, WB, LC-MS/MS	[59, 63]
IGFBP-2 and anti-IGFBP-2 autoantibodies	Lung cancer	ELISA, IHC	[64]
PRX1	Lung cancer	ELISA	[65]
s-endoglin	NSCLC	ELISA	[59, 66]
Pgrmc1	Lung cancer	WB	[59, 67]
proGRP (residue 31–98)	Lung cancer	LC/SRM/MS	[59, 68]
Ciz1b	Lung cancer	IF	[59]
MMP-1	Lung cancer	ELISA	[59, 70]
Cleaved uPAR	NSCLC	ELISA	[71]
ADAM28	NSCLC	ELISA	[72]
ALDH1A1	NSCLC	ELISA	[72]

Table 2.
New potential lung cancer serological biomarkers.

of the vascular endothelium. The main function of CD105 is to help the binding of endothelial cells to integrins and other receptors promoting angiogenesis by the activation of endothelial cells. CD105 overexpression was found in the endothelium of vessels in human solid tumors and is closely associated with a poor prognosis and the presence of metastases. Moreover levels of soluble CD105 (s-endoglin), formed by the cleavage of ectodomain of membrane receptors, were higher in patients with various types of cancer compared to normal counterparts, and its levels were also associated to metastases. In NSCLC s-endoglin serum levels were significantly decreased in postoperation patients, confirming its potential use for monitoring and prognosis of lung cancer [59, 66]. Progesterone receptor membrane component 1 (Pgrmc1) is a cytochrome b5-related protein induced by carcinogens. In fact, Pgrmc1 levels are elevated in spontaneous ovarian, breast, and lung cancers. Pgrmc1 is known to localize at the endoplasmic reticulum, and it was identified as a sigma-2 receptor, which is induced in cancers. Recently, Pgrmc1 showed also the potential to be a serum biomarker for lung cancer. It was shown that Pgrmc1 is localized in secretory vesicles and is secreted by lung cancer cells. Moreover, Pgrmc1 levels in the plasma and in the exosome fractions of plasma were significantly increased in lung cancer patients [59, 67]. Pro-gastrin-releasing peptide (proGRP, residue 31–98) is a more stable biochemical precursor of gastrin-releasing peptide (GRP), which is specifically produced by the neuroendocrine origin of SCLC cells. In a recent report, proGRP levels were increased in SCLC patients with respect to patients with other types of lung cancer; its levels are also associated with the progression of the disease [59, 68]. Ciz1 is a nuclear matrix protein which promotes the initiation of mammalian DNA replication. Recently, variant Ciz1 (24 nucleotides from the 3′end of exon 14 are excluded, leading to in frame deletion of eight amino acids 'VEEELCKQ') protein levels were significantly increased in the plasma of early stage lung cancer patients compared with that from healthy donors and with other respiratory diseases suggesting its potential use as a diagnostic lung cancer biomarker. Its sensitivity and specificity for stage I NSCLC were 95 and 74%, respectively [59]. MMP-1 is a collagenase that cleaves collagen types I, II, III, IV, and X at one site in the helical structure and is overexpressed in various cancer cells. High plasma levels of MMP-1 seem to be associated with a lower patient survival rate [59, 69]. uPAR is a glycosylphosphatidylinositol (GPI)-anchored glycoprotein and cell surface receptor specific to the urokinase plasminogen activator (uPA). The uPA-catalyzed cleavage of uPAR is a negative feedback loop in which uPA cleaves uPAR leaving the cleaved form of uPAR attached to the cell surface. It was shown that serum levels in preoperative NSCLC patients are correlated with higher levels of cleaved uPAR and lower survival rate [59, 70]. Kuroda et al. showed that ADAM28, a disintegrin and metalloproteinase 28 overexpressed in NSCLC tissues, was detectable also in serum of patients and increases with progress of tumor stage [71]. The sensitivity, false-negative rate, and AUC for ADAM28 were even better than those for CEA, suggesting a potential use of this test for diagnosis and monitoring of NSCLC. Recently, serum levels of ALDH1A1 were shown to be elevated in the sera of patients with NSCLC. Combined testing of serum ALDH1A1 and CEA levels significantly increased the screening sensitivity of CEA alone [72]. We provided evidence that isoforms other than ALDH1A1 may be secreted into the blood of lung cancer patients, and therefore screening sensitivity may be further enhanced by using an isoform-unspecific ALDH test without apparently affecting specificity [unpublished results]. Our results showed elevated ALDH serum levels can be detected in the vast majority of patients with early and advanced stage disease, suggesting that serum ALDH should be evaluated as part of a marker panel for noninvasive detection of early lung cancer in a larger cohort of patients at risk.

Although identification of proteins is now promising, quantification of proteins in complicated mixtures by MS remains challenging, especially in plasma carrying a large amount of proteins. Using antibody-based techniques such as ELISA, for biomarker measurements, could be hindered by a lack of high-quality antibodies. A quantitative approach has evolved, which performs targeted analysis of representative peptides by multiple reaction monitoring (MRM), and evaluated the potential utility of a list of candidate proteins for lung cancer diagnosis.

4.2 Liquid biopsy

Current methods employed for the evaluation of cancer genomes require tissue biopsy, either bone marrow biopsy of blood cancer or biopsy of affected nodal/soft tissue. These procedures are invasive and are limited by being representative of only a single site of the disease under evaluation. Thus, single-site tissue biopsy may not truly reflect the entire mutational profile of an individual's disease, and from a logistical perspective, it may not be suited to repeat biopsy over short periods of time. Liquid biopsy seems to be the most promising and, in contrast to tissue biopsy, is noninvasive, can provide a better representation of cancer genetic profile, and can be easily repeated [73] (**Figure 1**). A liquid biopsy is a blood sample of about 10–20 ml taken for diagnosis, prognosis, and prediction of a treatment purposes. It is a noninvasive approach to screening and early diagnosis of lung cancer. It consists not only of biomarkers but also circulating free DNA (cfDNA) and circulating tumor cells

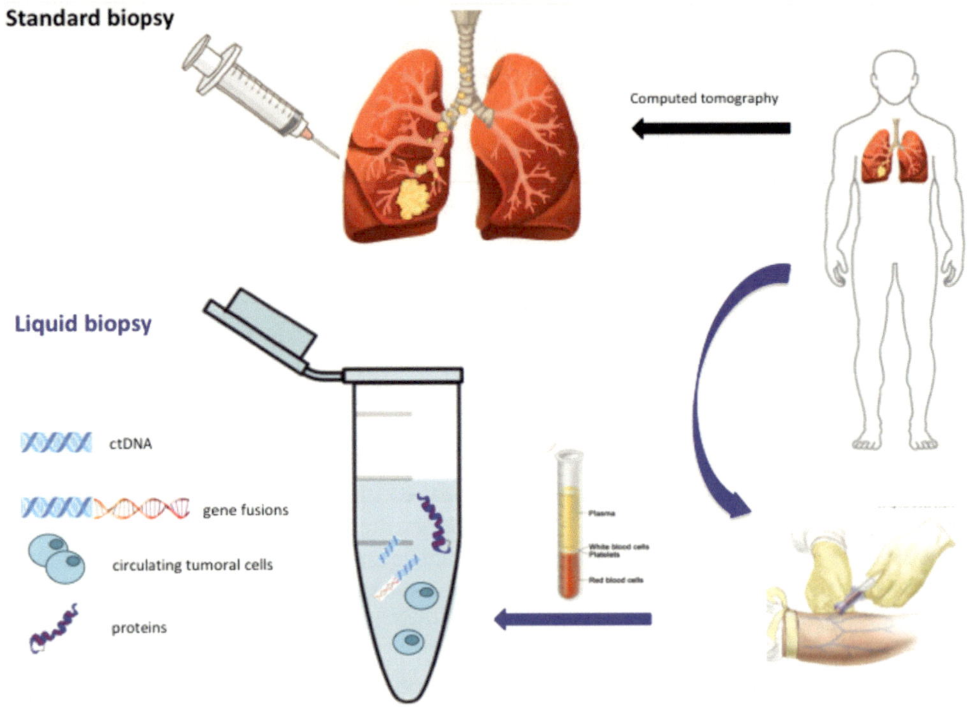

Figure 1.
Standard biopsy versus liquid biopsy. The figure illustrates the steps of the standard biopsy, beginning from the computed tomography for detecting pulmonary nodules. These are subsequently analyzed by extracting cells or tissue fragments, which are examined to determine the presence of the tumor. Liquid biopsy is a 10–20 ml blood sample taken for diagnosis, from which CTCs, biomarkers, and cfDNA—including circulating tumoral DNA (ctDNA)—are isolated and then analyzed in laboratory.

(CTCs). There are two methods for CTC isolation: the indirect CellSearch and the direct isolation by size of epithelial tumor cells (ISET) filtration method, which is, to date, the most interesting method for early diagnosis and screening of lung cancer using CTCs. CTCs can be cytomorphologically characterized before surgery and can be detected from asymptomatic patients with stage I lung cancer. Moreover, recent studies showed that CTCs can be isolated from patients with high risk of developing lung cancer (smokers with chronic obstructive pulmonary disease) without nodules detected by CT and that at the follow-up resulted positive for Ade [74, 75]. It was also showed that the initial presence of CTCs had a predictive value of 100% of developing a secondary lung cancer. As a follow-up to this pilot study performed in a single center and on a restricted number of patients, a multicenter study (named AIR project) began and has been involving 20 French university hospitals and 600 patients with chronic obstructive disease, over 55 years who smoked more than 30 packets per year. This project aims to study patients by means of CTC detection through ISET along with CT. Nucleic acids as DNA or RNA fragments can circulate in the plasma either freely or present in vesicle, as exosomes. While cfDNA is universally found in the plasma of healthy people as well as those with benign diseases, it has been observed that patients with malignant disease have higher levels of cfDNA in their plasma [76]. Among RNA, coding (microRNA) and noncoding RNA can circulate. Recent studies have evidenced the more or less complex signature of plasma microRNA associated with lung cancer. In particular, it was showed that a signature of several plasma microRNA has a predictive value for lung cancer in a high-risk population [74]. Although liquid biopsy can permit the monitoring of patients on treatment or after treatment for lung cancer, it holds some limits. The major limits for the use of circulating nucleic acids for early lung cancer detection concerns the distinction between free nucleic acid of germinal or somatic origin because of the lysis of circulating hematological cells and also the low amount detectable in very early tumor stages. Moreover, circulating somatic microRNA can be released from other diseases associated with lung cancer, such as cardiovascular diseases, inflammatory disorders, pulmonary fibrosis, and other associated cancers. Another limit is relative to the pre-analytical phase, which is very crucial as the delay between blood sampling and the analytical phase should be as short as possible. It should also be the same for all the patients involved in the study, for a better comparison of the results, and the blood must be conditioned with a buffer that allows excellent conservation of the material. Moreover, the low amount of biomarkers, cfDNA or CTCs in the blood of patients with a very small tumor or not still visible by thoracic imagery, needs a high sensitive technique, which should be able to isolate enough material to conduct the analysis. Another problem is represented by the lack of standardization of the different pre-analytical and analytical steps which limit the deployment of liquid biopsy in clinical practice for early diagnosis of lung cancer [77–79]. Therefore, it is difficult to obtain a specific result if the pre-analytical phase is not standardized among all the patients involved in the study and, also, if the clinico-biological information about the patient is not known, as other potential comorbidities can emerge.

Other than variants in cfDNA, aberrant DNA methylation of some novel and known genes was also investigated in serum of patients with lung cancer by means of a quantitative methylation-specific PCR and showed a specificity of 71% [77]. Identification of blood-based noninvasive or minimally invasive detection markers will improve the clinical management of lung cancer. It is noteworthy that this simple, reliable, and noninvasive blood test could aid not only the early detection of lung cancer but also could be used most effectively to direct imaging modalities with low specificity such as CT. Such a test could therefore have a significant impact on the long-term survival of these individuals. Moreover, these tests can be helpful in monitoring the response to therapy and to identify new actionable mutations.

5. Conclusions

This review summarizes the molecular pathology and the conventional methods used for screening of lung cancer, highlighting the advantages and limits of these approaches. We also report the recent studies about new circulating biomarkers potentially useful for lung cancer screening. Ideally, a biomarker should have a sensitivity and specificity of 100%, a goal that is almost never achieved. One strategy potentially increasing both parameters is to combine several biomarkers into a screening marker panel. Combined with other noninvasive methods, this may allow for further refinement of lung cancer screening. Liquid biopsy is a 10 ml blood sample taken for diagnostic, prognostic, and disease monitoring purposes. It consists not only of biomarkers but also circulating cfDNA, RNA, and CTCs. With respect to tissue biopsy, it permits to have a better representation of the whole cancer genetic profile and may be suited to repeat biopsy over short periods of time. Compared to imaging modalities such as X-rays and CT, liquid biopsy represents a more reliable, less invasive, and less expensive method for the detection of lung cancer in populations at risk of developing this disease.

Acknowledgements

The study was supported by the T. & L. de "Beaumont Bonelli Foundation for cancer research" (to AR). We are thankful to Dr. Stephen Duckworth for his support in editing of the manuscript, the Committee on Biotechnologies and VirusSphere, World Academy of Biomedical Technologies, UNESCO, Paris, France (to GT), and Thomas More University U.P.T.M., Rome, Italy (to GT).

Conflict of interests

The authors declare no conflicts of interest.

Author contributions

All authors made substantial contributions to each stage of the preparation of this manuscript for publication. All authors approved the final version.

Funding

This work was supported by T. & L. de Beaumont Bonelli Foundation for cancer research.

Author details

Giulio Tarro[1], Moreno Paolini[2] and Alessandra Rossi[2]*

1 T. & L. de Beaumont Bonelli Foundation, Naples, Italy

2 Department of Pharmacy and Biotechnologies, Alma Mater Studiorum, University of Bologna, Bologna, Italy

*Address all correspondence to: alessandra.rossi10@studio.unibo.it

References

[1] Siegel R, Naishadham D, Jemal A. Cancer statistics, 2013. CA: A Cancer Journal for Clinicians. 2013;**63**(1):11-30

[2] Jemal A, Bray F, Center MM, Ferlay J, Ward E, Forman D. Global cancer statistics. CA: A Cancer Journal for Clinicians. 2011;**61**(2):69-90

[3] Yuxia M, Zhennan T, Wei Z. Circulating miR-125b is a novel biomarker for screening non-small-cell lung cancer and predicts poor prognosis. Journal of Cancer Research and Clinical Oncology. 2012;**138**(12): 2045-2050

[4] Zhao S, Jiang T, Zhang L, Yang H, Liu X, Jia Y, et al. Clinicopathological and prognostic significance of regulatory T cells in patients with non-small cell lung cancer: A systematic review with meta-analysis. Oncotarget. 2016;7(24):36065-36073

[5] Reinersman JM, Johnson ML, Riely GJ, Chitale DA, Nicastri AD, Soff GA, et al. Frequency of EGFR and KRAS mutations in lung adenocarcinomas in African Americans. Journal of Thoracic Oncology. 2011;**6**(1):28-31

[6] Hudson AM, Wirth C, Stephenson NL, Fawdar S, Brognard J, Miller CJ. Using large-scale genomics data to identify driver mutations in lung cancer: Methods and challenges. Pharmacogenomics. 2015;**16**(10):1149-1160

[7] Makinoshima H, Umemura S, Suzuki A, Nakanishi H, Maruyama A, Udagawa H, et al. Metabolic determinants of sensitivity to phosphatidylinositol 3-kinase pathway inhibitor in small-cell lung carcinoma. Cancer Research. 2018;**78**(9):2179-2190

[8] Fernandez-Cuesta L, Plenker D, Osada H, Sun R, Menon R, Leenders F, et al. CD74-NRG1 fusions in lung adenocarcinoma. Cancer Discovery. 2014;**4**(4):415-422

[9] Rikova K, Guo A, Zeng Q, Possemato A, Yu J, Haack H, et al. Global survey of phosphotyrosine signaling identifies oncogenic kinases in lung cancer. Cell. 2007;**131**(6):1190-1203

[10] Soda M, Choi YL, Enomoto M, Takada S, Yamashita Y, Ishikawa S, et al. Identification of the transforming EML4-ALK fusion gene in non-small-cell lung cancer. Nature. 2007;**448**(7153):561-566

[11] Stephens P, Hunter C, Bignell G, Edkins S, Davies H, Teague J, et al. Lung cancer: Intragenic ERBB2 kinase mutations in tumours. Nature. 2004;**431**(7008):525-526

[12] Chen Z, Fillmore CM, Hammerman PS, Kim CF, Wong KK. Non-small-cell lung cancers: A heterogeneous set of diseases. Nature Reviews. Cancer. 2014;**14**(8):535-546

[13] Vaishnavi A, Capelletti M, Le AT, Kako S, Butaney M, Ercan D, et al. Oncogenic and drug-sensitive NTRK1 rearrangements in lung cancer. Nature Medicine. 2013;**19**(11):1469-1472

[14] Weiss J, Sos ML, Seidel D, Peifer M, Zander T, Heuckmann JM, et al. Frequent and focal FGFR1 amplification associates with therapeutically tractable FGFR1 dependency in squamous cell lung cancer. Science Translational Medicine. 2010;**2**(62):62-93

[15] Guagnano V, Kauffmann A, Wöhrle S, Stamm C, Ito M, Barys L, et al. FGFR genetic alterations predict for sensitivity to NVP-BGJ398, a selective pan-FGFR inhibitor. Cancer Discovery. 2012;**2**(12):1118-1133

[16] Ricordel C, Lespagnol A, Llamas-Gutierrez F, de Tayrac M, Kerjouan M,

Fievet A, et al. Mutational landscape of DDR2 gene in lung squamous cell carcinoma using next-generation sequencing. Clinical Lung Cancer. 2018;**19**(2):163-169 e4

[17] Al-Hajj M, Wicha MS, Benito-Hernandez A, Morrison SJ, Clarke MF. Prospective identification of tumorigenic breast cancer cells. Proceedings of the National Academy of Sciences of the United States of America. 2003;**100**(7):3983-3988

[18] Singh SK, Clarke ID, Terasaki M, Bonn VE, Hawkins C, Squire J, et al. Identification of a cancer stem cell in human brain tumors. Cancer Research. 2003;**63**(18):5821-5828

[19] Ricci-Vitiani L, Lombardi DG, Pilozzi E, Biffoni M, Todaro M, Peschle C, et al. Identification and expansion of human colon-cancer-initiating cells. Nature. 2007;**445**(7123):111-115

[20] Kelly K, Huang C. Biological agents in non-small cell lung cancer: A review of recent advances and clinical results with a focus on epidermal growth factor receptor and vascular endothelial growth factor. Journal of Thoracic Oncology. 2008;**3**(6):664-673

[21] Jelski W, Chrostek L, Markiewicz W, Szmitkowski M. Activity of alcohol dehydrogenase (ADH) isoenzymes and aldehyde dehydrogenase (ALDH) in the sera of patients with breast cancer. Journal of Clinical Laboratory Analysis. 2006;**20**(3):105-108

[22] Guo X, Wang Y, Lu H, Cai X, Wang X, Zhou Z, et al. Genome-wide characterization and expression analysis of the aldehyde dehydrogenase (ALDH) gene superfamily under abiotic stresses in cotton. Gene. 2017;**628**:230-245

[23] Meng E, Mitra A, Tripathi K, Finan MA, Scalici J, McClellan S, et al. ALDH1A1 maintains ovarian cancer stem cell-like properties by altered regulation of cell cycle checkpoint and DNA repair network signaling. PLoS One. 2014;**9**(9):e107142

[24] Zhou L, Sheng D, Wang D, Ma W, Deng Q, Deng L, et al. Identification of cancer-type specific expression patterns for active aldehyde dehydrogenase (ALDH) isoforms in ALDEFLUOR assay. Cell Biology and Toxicology. 2018

[25] Li J, Zhang B, Yang YF, Jin J, Liu YH. Aldehyde dehydrogenase 1 as a predictor of the neoadjuvant chemotherapy response in breast cancer: A meta-analysis. Medicine (Baltimore). 2018;**97**(34):e12056

[26] Liang D, Shi Y. Aldehyde dehydrogenase-1 is a specific marker for stem cells in human lung adenocarcinoma. Medical Oncology. 2012;**29**(2):633-639

[27] López-González A, Salas C, Provencio M, Córdoba M, Gamallo C. Aldehyde dehydrogenases in early stage lung cancer: Nuclear expression. Clinical & Translational Oncology. 2014;**16**(10):931-934

[28] Qiu Y, Pu T, Li L, Cheng F, Lu C, Sun L, et al. The expression of aldehyde dehydrogenase family in breast cancer. Journal of Breast Cancer. 2014;**17**(1):54-60

[29] Ran D, Schubert M, Taubert I, Eckstein V, Bellos F, Jauch A, et al. Heterogeneity of leukemia stem cell candidates at diagnosis of acute myeloid leukemia and their clinical significance. Experimental Hematology. 2012;**40**(2):155-65.e1

[30] Xing Y, Luo DY, Long MY, Zeng SL, Li HH. High ALDH1A1 expression correlates with poor survival in papillary thyroid carcinoma. World Journal of Surgical Oncology. 2014;**12**:29

[31] Qian X, Wagner S, Ma C, Coordes A, Gekeler J, Klussmann JP, et al.

Prognostic significance of ALDH1A1-positive cancer stem cells in patients with locally advanced, metastasized head and neck squamous cell carcinoma. Journal of Cancer Research and Clinical Oncology. 2014;**140**(7):1151-1158

[32] Miyata T, Oyama T, Yoshimatsu T, Higa H, Kawano D, Sekimura A, et al. The clinical significance of cancer stem cell markers ALDH1A1 and CD133 in lung adenocarcinoma. Anticancer Research. 2017;**37**(5):2541-2547

[33] Gao F, Zhou B, Xu JC, Gao X, Li SX, Zhu GC, et al. The role of LGR5 and ALDH1A1 in non-small cell lung cancer: Cancer progression and prognosis. Biochemical and Biophysical Research Communications. 2015;**462**(2):91-98

[34] MacDonagh L, Gallagher MF, Ffrench B, Gasch C, Breen E, Gray SG, et al. Targeting the cancer stem cell marker, aldehyde dehydrogenase 1, to circumvent cisplatin resistance in NSCLC. Oncotarget. 2017;**8**(42):72544-72563

[35] Zhou Y, Wang Y, Ju X, Lan J, Zou H, Li S, et al. Clinicopathological significance of ALDH1A1 in lung, colorectal, and breast cancers: A meta-analysis. Biomarkers in Medicine. 2015;**9**(8):777-790

[36] Patel M, Lu L, Zander DS, Sreerama L, Coco D, Moreb JS. ALDH1A1 and ALDH3A1 expression in lung cancers: Correlation with histologic type and potential precursors. Lung Cancer. 2008;**59**(3):340-349

[37] Moreb JS, Baker HV, Chang LJ, Amaya M, Lopez MC, Ostmark B, et al. ALDH isozymes downregulation affects cell growth, cell motility and gene expression in lung cancer cells. Molecular Cancer. 2008;7:87

[38] Marchitti SA, Orlicky DJ, Brocker C, Vasiliou V. Aldehyde dehydrogenase 3B1 (ALDH3B1): Immunohistochemical tissue distribution and cellular-specific localization in normal and cancerous human tissues. The Journal of Histochemistry and Cytochemistry. 2010;**58**(9):765-783

[39] Aberle DR, DeMello S, Berg CD, Black WC, Brewer B, Church TR, et al. Results of the two incidence screenings in the National Lung Screening Trial. The New England Journal of Medicine. 2013;**369**(10):920-931

[40] Henschke CI, Yip R, Yankelevitz DF, Smith JP, I.E.L.C.A.P. Investigators. Definition of a positive test result in computed tomography screening for lung cancer: A cohort study. Annals of Internal Medicine. 2013;**158**(4):246-252

[41] Infante M, Cavuto S, Lutman FR, Passera E, Chiarenza M, Chiesa G, et al. Long-term follow-up results of the DANTE trial, a randomized study of lung cancer screening with spiral computed tomography. American Journal of Respiratory and Critical Care Medicine. 2015;**191**(10):1166-1175

[42] Pastorino U, Rossi M, Rosato V, Marchianò A, Sverzellati N, Morosi C, et al. Annual or biennial CT screening versus observation in heavy smokers: 5-year results of the MILD trial. European Journal of Cancer Prevention. 2012;**21**(3):308-315

[43] Wille MM, Dirksen A, Ashraf H, Saghir Z, Bach KS, Brodersen J, et al. Results of the randomized Danish lung cancer screening trial with focus on high-risk profiling. American Journal of Respiratory and Critical Care Medicine. 2016;**193**(5):542-551

[44] Berlin NI, Buncher CR, Fontana RS, Frost JK, Melamed MR. The National Cancer Institute cooperative early lung cancer detection program. Results of the initial screen (prevalence). Early lung cancer detection: Introduction. The American Review of Respiratory Disease. 1984;**130**(4):545-549

[45] Marcus PM, Bergstralh EJ, Zweig MH, Harris A, Offord KP, Fontana RS. Extended lung cancer incidence follow-up in the Mayo lung project and overdiagnosis. Journal of the National Cancer Institute. 2006;**98**(11):748-756

[46] Rhim AD, Mirek ET, Aiello NM, Maitra A, Bailey JM, McAllister F, et al. EMT and dissemination precede pancreatic tumor formation. Cell. 2012;**148**(1-2):349-361

[47] Sone S, Li F, Yang ZG, Honda T, Maruyama Y, Takashima S, et al. Results of three-year mass screening programme for lung cancer using mobile low-dose spiral computed tomography scanner. British Journal of Cancer. 2001;**84**(1):25-32

[48] Lyne C, Zaw S, King B, See K, Manners D, Al-Kaisey A, et al. Low rates of eligibility for lung cancer screening in patients undergoing computed tomography coronary angiography. Internal Medicine Journal. 2018;**48**(10):1265-1268

[49] Cai J, Xu D, Liu S, Cham MD. The added value of computer-aided detection of small pulmonary nodules and missed lung cancers. Journal of Thoracic Imaging. 2018;**33**(6):390-395

[50] Reich JM. A critical appraisal of overdiagnosis: Estimates of its magnitude and implications for lung cancer screening. Thorax. 2008;**63**(4):377-383

[51] Schneider J. Tumor markers in detection of lung cancer. Advances in Clinical Chemistry. 2006;**42**:1-41

[52] Duffy MJ, O'Byrne K. Tissue and blood biomarkers in lung cancer: A review. Advances in Clinical Chemistry. 2018;**86**:1-21

[53] Ferrigno D, Buccheri G, Giordano C. Neuron-specific enolase is an effective tumour marker in non-small cell lung cancer (NSCLC). Lung Cancer. 2003;**41**(3):311-320

[54] Liu J, Zhu H, Jiang H, Zhang H, Wu D, Hu X, et al. Tumor M2 pyruvate kinase in diagnosis of nonsmall cell lung cancer: A meta-analysis based on Chinese population. Journal of Cancer Research and Therapeutics. 2015;**11**(Suppl 1):C104-C106

[55] Xu M, Zhu M, Du Y, Yan B, Wang Q, Wang C, et al. Serum C-reactive protein and risk of lung cancer: A case-control study. Medical Oncology. 2013;**30**(1):319

[56] Leidinger P, Keller A, Heisel S, Ludwig N, Rheinheimer S, Klein V, et al. Identification of lung cancer with high sensitivity and specificity by blood testing. Respiratory Research. 2010;**11**:18

[57] Wu HY, Goan YG, Chang YH, Yang YF, Chang HJ, Cheng PN, et al. Qualification and verification of serological biomarker candidates for lung adenocarcinoma by targeted mass spectrometry. Journal of Proteome Research. 2015;**14**(8): 3039-3050

[58] Begum S, Brait M, Dasgupta S, Ostrow KL, Zahurak M, Carvalho AL, et al. An epigenetic marker panel for detection of lung cancer using cell-free serum DNA. Clinical Cancer Research. 2011;**17**(13):4494-4503

[59] Heo SH, Lee SJ, Ryoo HM, Park JY, Cho JY. Identification of putative serum glycoprotein biomarkers for human lung adenocarcinoma by multilectin affinity chromatography and LC-MS/MS. Proteomics. 2007;**7**(23):4292-4302

[60] Chee J, Naran A, Misso NL, Thompson PJ, Bhoola KD. Expression of tissue and plasma kallikreins and kinin B1 and B2 receptors in lung cancer. Biological Chemistry. 2008;**389**(9):1225-1233

[61] Sung HJ, Jeon SA, Ahn JM, Seul KJ, Kim JY, Lee JY, et al. Large-scale isotype-specific quantification of serum amyloid a 1/2 by multiple reaction monitoring in crude sera. Journal of Proteomics. 2012;**75**(7):2170-2180

[62] Kang SM, Sung HJ, Ahn JM, Park JY, Lee SY, Park CS, et al. The Haptoglobin β chain as a supportive biomarker for human lung cancers. Molecular BioSystems. 2011;**7**(4):1167-1175

[63] Narayanasamy A, Ahn JM, Sung HJ, Kong DH, Ha KS, Lee SY, et al. Fucosylated glycoproteomic approach to identify a complement component 9 associated with squamous cell lung cancer (SQLC). Journal of Proteomics. 2011;**74**(12):2948-2958

[64] Zhang Y, Ying X, Han S, Wang J, Zhou X, Bai E, et al. Autoantibodies against insulin-like growth factor-binding protein-2 as a serological biomarker in the diagnosis of lung cancer. International Journal of Oncology. 2013;**42**(1):93-100

[65] Rostila A, Puustinen A, Toljamo T, Vuopala K, Lindström I, Nyman TA, et al. Peroxiredoxins and tropomyosins as plasma biomarkers for lung cancer and asbestos exposure. Lung Cancer. 2012;**77**(2):450-459

[66] Kopczyńska E, Dancewicz M, Kowalewski J, Makarewicz R, Kardymowicz H, Kaczmarczyk A, et al. Time-dependent changes of plasma concentrations of angiopoietins, vascular endothelial growth factor, and soluble forms of their receptors in nonsmall cell lung cancer patients following surgical resection. ISRN Oncology. 2012;**2012**:638352

[67] Mir SU, Ahmed IS, Arnold S, Craven RJ. Elevated progesterone receptor membrane component 1/ sigma-2 receptor levels in lung tumors and plasma from lung cancer patients. International Journal of Cancer. 2012;**131**(2):E1-E9

[68] Wu XY, Hu YB, Li HJ, Wan B, Zhang CX, Zhang B, et al. Diagnostic and therapeutic value of progastrin-releasing peptide on small-cell lung cancer: A single-center experience in China. Journal of Cellular and Molecular Medicine. 2018;**22**(9):4328-4334

[69] Selvaraj G, Kaliamurthi S, Lin S, Gu K, Wei DQ. Prognostic impact of tissue inhibitor of metalloproteinase-1 in non-small cell lung cancer: Systematic review and meta-analysis. Current Medicinal Chemistry. 2018. In press

[70] Almasi CE, Drivsholm L, Pappot H, Hoyer-Hansen G, Christensen IJ. The liberated domain I of urokinase plasminogen activator receptor—A new tumour marker in small cell lung cancer. APMIS. 2013;**121**(3):189-196

[71] Kuroda H, Mochizuki S, Shimoda M, Chijiiwa M, Kamiya K, Izumi Y, et al. ADAM28 is a serological and histochemical marker for non-small-cell lung cancers. International Journal of Cancer. 2010;**127**(8):1844-1856

[72] Cao YT, Li JH, Wang YT, Fu YW, Xu J. Serum ALDH1A1 is a tumor marker for the diagnosis of non-small cell lung cancer. Tumori. 2014;**100**(2):214-218

[73] Ilié M, Hofman P. Pros: Can tissue biopsy be replaced by liquid biopsy? Translational Lung Cancer Research. 2016;**5**(4):420-423

[74] Ilie M, Hofman V, Long-Mira E, Selva E, Vignaud JM, Padovani B, et al. "Sentinel" circulating tumor cells allow early diagnosis of lung cancer in patients with chronic obstructive pulmonary disease. PLoS One. 2014;**9**(10):e111597

[75] Lam VK, Tran HT, Banks KC, Lanman RB, Rinsurongkawong W, Peled N, et al. Targeted tissue and cell-free tumor DNA sequencing of advanced

lung squamous-cell carcinoma reveals clinically significant prevalence of actionable alterations. Clinical Lung Cancer. 2018;**20**(1):30-36

[76] Wei L, Wu W, Han L, Yu W, Du Y. A quantitative analysis of the potential biomarkers of non-small cell lung cancer by circulating cell-free DNA. Oncology Letters. 2018;**16**(4):4353-4360

[77] Hofman VJ, Ilie M, Hofman PM. Detection and characterization of circulating tumor cells in lung cancer: Why and how? Cancer Cytopathology. 2016;**124**(6):380-387

[78] Ilie M, Hofman V, Long E, Bordone O, Selva E, Washetine K, et al. Current challenges for detection of circulating tumor cells and cell-free circulating nucleic acids, and their characterization in non-small cell lung carcinoma patients. What is the best blood substrate for personalized medicine? Annals of Translational Medicine. 2014;**2**(11):107

[79] Montani F, Marzi MJ, Dezi F, Dama E, Carletti RM, Bonizzi G, et al. miR-test: A blood test for lung cancer early detection. Journal of the National Cancer Institute. 2015;**107**(6):djv063

[80] Tran TN, Selinger CI, Yu B, Ng CC, Kohonen-Corish MR, McCaughan B, et al. Alterations of insulin-like growth factor-1 receptor gene copy number and protein expression are common in non-small cell lung cancer. Journal of Clinical Pathology. 2014;**67**(11):985-991